High Frontier

The U. S. Air Force
and
the Military Space Program

Curtis Peebles

Air Force History and Museums Program
1997

TABLE OF CONTENTS

Page

The USAF and the Military Space Program

The United States military space program began at the end of World War II when a few people realized that space flight was now achievable and could be employed to military advantage. Science and technology in the form of advanced radar, jet propulsion, ballistic rockets such as the V-2, and nuclear energy had dramatically altered the nature of war. Army Air Forces Commanding General Henry H. "Hap" Arnold wrote in November 1945 that a space ship "is all but practicable today" and could be built "within the foreseeable future." The following month the Air Force Scientific Advisory Group concluded that long-range rockets were technically feasible and that satellites were a "definite possibility." The U.S. Navy also expressed interest in space flight. In November 1945 the Navy Bureau of Aeronautics produced a satellite report, and, on March 7, 1946 proposed an interservice space program. The idea was presented to the joint Army-Navy Aeronautical Board on April 9. Major General Curtis E. LeMay, the Director of Research and Development for the Army Air Forces, however, viewed space operations as an exclusive Air Force domain, and he ordered an independent study.

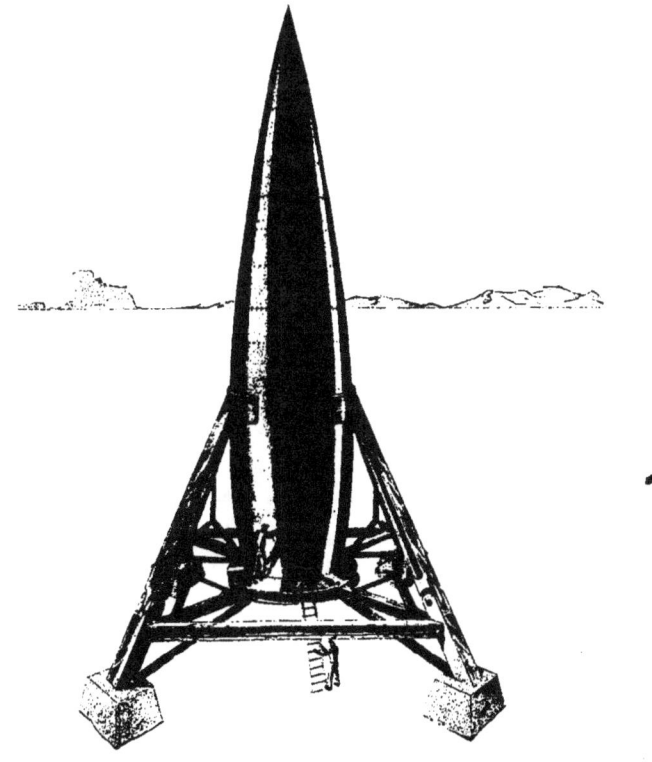

1946 RAND Satellite Rocket

I

Military Space Missions Defined

To conduct this study, the Army Air Forces turned to Project RAND, then a section of Douglas Aircraft Company, established to provide long-range technical advice to the service. The resultant study, "Preliminary Design of an Experimental World-Circling Spaceship" was issued on May 2, 1946, and embraced a wide-ranging examination of satellite technology. A "satellite offers an observation aircraft which cannot be brought down by an enemy . . . ," the report observed. Other military roles included the "spotting of the points of impact of bombs launched by us, and the observation of weather conditions over enemy territory." The report also mentioned satellites used for communications, and addressed other, as yet unforeseen potentialities:

> In making the decision as to whether or not to undertake construction of such a craft now, it is not inappropriate to view our present situation as similar to that in airplanes prior to the flight of the Wright brothers. We can see no more clearly all the utility and implications of spaceships than the Wright brothers could see fleets of B–29s bombing Japan and air transports circling the globe.

Other studies followed, and in December 1947 the Air Materiel Command (AMC) completed a review of the RAND studies. AMC confirmed that satellites were technically feasible, but proposed further evaluation of satellite requirements and specifications. A few weeks later, on January 15, 1948, General Hoyt S. Vandenberg claimed for the newly-created Air Force "logical responsibility for the satellite."

The uneasy alliance among wartime Allies broke down during 1947 and 1948, when the Soviet Union installed Communist governments throughout Eastern Europe. In Western Europe, Communist parties sought to extend Moscow's control, while in Greece and Turkey, communist insurgents battled Western-backed governments. In June 1948, the Soviets sealed off all land routes into West Berlin. For nearly a year, the city had to be supplied by air. Then, in August 1949, the Soviets tested their first atomic bomb, bringing to an end the U.S. nuclear monopoly sooner than had been expected. A few months later, communist forces swept to power in China. Then, on June 25, 1950, North Korean troops invaded South Korea, and within days U.S. forces were committed under the United Nations. In November, Communist Chinese "volunteers" entered the Korean War.

All the while, the Soviets' ability to launch nuclear attacks against the United States increased. In August 1953 the Soviets tested their first hydrogen bomb, and, at the 1954 May Day parade in Moscow, unveiled a single Mya 4 jet bomber. Armed with hydrogen bombs, such aircraft posed a major threat to America. During the July 13, 1955 Aviation Day parade, Mya 4s made a mass fly-by. Estimates that the Soviets were building a bomber force larger than that of the U.S. began to appear, and, with them, a controversy over an American "Bomber Gap." If almost nothing was known about Soviet nuclear capabilities, its military strengths and weaknesses, or intentions, the first attempts to breach this secrecy were already underway.

Early Aerial Reconnaissance

At the start of the Cold War, the United States began aerial reconnaissance about the periphery of the Soviet Union. One major effort involved detecting Soviet nuclear tests. Air Force weather planes, fitted with air scoops to collect radioactive fallout in 1949 returned the first evidence that the Soviets had tested an atomic bomb. This early nuclear detonation detection program anticipated important Air Force space missions. Another effort, Project Mogul, used balloon-borne acoustic sensors in an attempt to detect Soviet nuclear and missile tests. These sensors, designed to pick up the low-frequency atmospheric shock waves radiated from the Soviet Union, were mounted in balloons flying over New Mexico. Debris from one of these balloons employed in training flights fell to Earth near Roswell, New Mexico, in early June 1947, prompting reports that a "flying saucer" had been recovered. Although Project Mogul had only limited success, it, too, anticipated real-time nuclear detection and ballistic missile early warning.

American reconnaissance flights about the Soviet Union's periphery yielded both photo and radar images of port facilities, coastal areas, and islands. But the vast interior of the Soviet Union remained out of reach. If the U.S. was to acquire reliable intelligence information of regions from which a Soviet surprise attack might be mounted, it would require deep overflights. In 1950, the best prospect for such deep overflights was camera-carrying balloons. Skyhook weather balloons could reach altitudes above the ceiling of jet fighters and ride the winter jet stream from west to east. Under the code name "Project Genetrix," the balloons would be launched in Europe and drift across the Soviet Union on prevailing winds. Once clear of Soviet airspace, the camera-carrying gondolas would be cut free by a radio command. Descending by parachute, they could be caught in midair by modified C–119 transports, or recovered on the ocean's surface.

On January 10, 1956, President Dwight D. Eisenhower gave his approval. Balloon launches began and continued over the next several weeks. On February 4, 1956, the Soviets vehemently protested the balloon flights. Two days later, Eisenhower ordered a halt to any further balloon launches. In all, 448 Genetrix balloons had been launched; of these, 40

Drawing of the Genetrix reconnaissance balloon payload. The Genetrix missions were flown in January and February 1956. Owing to extra ballast intended to lower the flight altitude, and Soviet air defenses, these balloons suffered a high loss rate and President Dwight Eisenhower terminated the program.

recovered gondolas returned photos. The 13,813 usable photos covered 1,116,449 square miles of the Soviet Union and China (eight percent of their total area). Because the balloons' path could not be controlled, however, much of the photography was of fields and forests, areas of little intelligence value.

A second series of reconnaissance balloon launches, known as the WS-461L program, began in July 1958. As designed, these balloons could fly for nearly a month at altitudes of 100,000 feet. But only three WS-461L reconnaissance balloons were launched from the aircraft carrier *USS Windham Bay* cruising south of Alaska. Because of a late start and an error in setting the timing release of the gondolas by the military launch crew, all three balloons descended to Earth inside Poland. An irate President Eisenhower ordered an end to all further reconnaissance balloon flights.

Fears of a Soviet surprise nuclear attack affected not just Western leaders. A poll conducted in the mid 1950s indicated that more than half of all American citizens thought that they were more likely to die in a nuclear attack than from old age. With each military service demanding funds to meet every possible contingency, President Eisenhower faced an intolerable situation. At a meeting with the Science Advisory Committee of the Office of Defense Mobilization on March 27, 1954, he said, "modern weapons had made it easier for a hostile nation with a closed society to plan an attack in secrecy and thus gain an advantage denied to the nation with an open society." He asked Lee A. DuBridge, President of the California Institute of Technology and Chairman of the Scientific Advisory Committee, and James R. Killian Jr., President of the Massachusetts Institute of Technology (MIT), to organize a special study group to consider continental defense, strike forces, and intelligence. Killian chaired this group, organized in July and first called the "Surprise Attack Panel," later the "Technological Capabilities Panel" (TCP). The TCP operated between August 1954 and January 1955. Its two-volume final report, "Meeting the Threat of Surprise Attack," was issued on February 14, 1955. A separate, Presidential annex prepared by Edwin Land, President of Polaroid, contained details of a special, unarmed, high-altitude reconnaissance aircraft that could overfly the Soviet Union.

Back in November 1954, at the urging of Land and Killian, President Eisenhower had approved development of the plane eventually known to history as the Lockheed U–2. Knowledge of the U–2 program, conducted by the Central Intelligence Agency (CIA) and the Air Force, was very closely held. Until May 1960 only four people in the White House and fewer than 400 in the country knew about this project. Eisenhower also made a political effort to secure international agreement permitting aerial reconnaissance. At a July 1955 Geneva Four Power Summit Conference, he proposed that reconnaissance aircraft of the U.S. and Soviet Union

An Air Force U-2 in flight. Between 1956 and 1960, CIA-operated U-2s made 24 overflights of the Soviet Union.

overfly each other's territory. The Soviets rejected this proposal, later known as "Open Skies," seeking to protect their closed society. Beginning in 1956, President Eisenhower personally approved each U–2 overflight, including the first one on the Fourth of July. Soon after the start of U–2 overflights, American civilian leaders *knew* that the Soviets were not building a large bomber force. The Soviets possessed about fifty Mya 4 jet bombers and Tu 95 turboprop bombers combined, far fewer than had been estimated and a far smaller number than the U.S. bomber force. For the first time, the U.S. possessed solid information on Soviet military power on which to base its own force structure. The U–2 heralded a major change in the acquisition of intelligence.

Military Space Missions Identified, Satellite Work Begun

The RAND satellite studies between 1947 and 1954 identified all of the military satellite missions that would form the future Air Force space program. RAND researchers Stanley M. Greenfield and William W. Kellogg authored a report titled, "Inquiry into the Feasibility of Weather Reconnaissance." The low resolution possible with television systems of the early 1950s was sufficient to identify cloud types. Other studies identified satellites used as communications relays and as platforms from which to guide aircraft, ships, and missiles. Relaying long-range communications with satellites would be particularly valuable in areas like the Pacific Ocean, or for communications between ships and shore stations. Given the Soviet threat, however, satellite reconnaissance claimed the primary interest. In April 1951, RAND issued a report, "Utility of a Satellite Vehicle for Reconnaissance." But reconnaissance satellites did not figure in the 1952 Beacon Hill study conducted by Project Lincoln for the Air Force, which looked to improve aerial and balloon reconnaissance systems. The members of the study apparently did not view satellite reconnaissance as technically feasible in the near term.

In May 1953, the Air Force Air Research and Development Command (ARDC) assumed "active direction" of RAND's satellite efforts, and staff members who visited RAND in August came away with the belief that an immediate satellite effort should be started even if a reconnaissance payload was not yet possible. On September 8, RAND recommended to ARDC that full system development begin "perhaps immediately following the completion of experimental component tests." That December, ARDC established Project 409-40, a "Satellite Component Study" for a reconnaissance satellite, later given the Air Force designation WS-117L. On March 1, 1954, RAND issued its final two-volume "Project Feed Back Summary Report." The report urged design, construction, and use of a "satellite reconnaissance vehicle" as a matter of "vital strategic interest to the United States." RAND's Feed Back study envisioned using an imaging orthicon television system. RAND engineers estimated that the satellite could provide 30 million pictures during a year of operation. As with

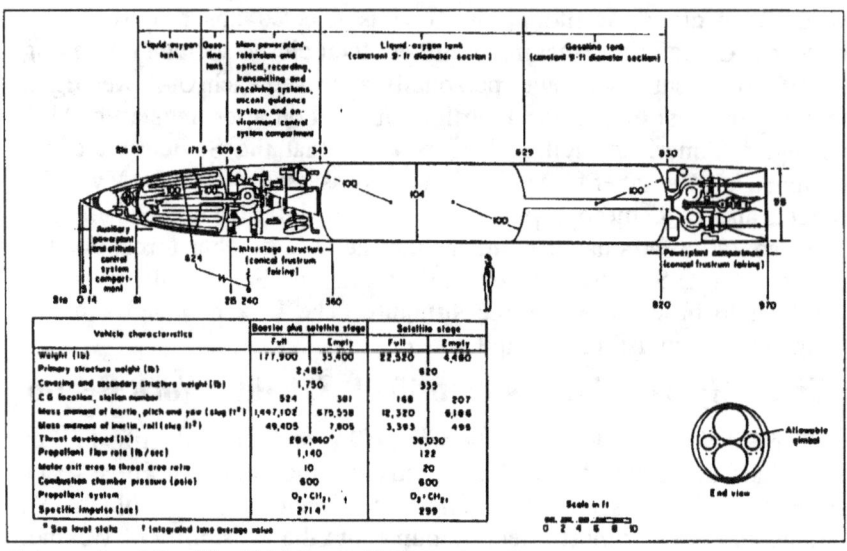

Schematic of Satellite Vehicle - RAND Project Feed Back, March 1954.

the earlier proposals, a low resolution at the surface—about 144 feet—remained the main stumbling block.

At this time, technical breakthroughs made possible the production of small nuclear warheads. In February 1954 a special Air Force evaluation panel, John von Neumann's Teapot Committee, recommended that the United States accelerate development of Intercontinental Ballistic Missiles (ICBMs). To direct this work, ARDC created the Western Development Division (WDD), with headquarters in Inglewood, California. The resultant ICBM, Air Force leaders knew, also could orbit significant payloads when fitted with powered upper stages. Up to this point, a separate satellite booster was planned. Using modified ballistic missiles as launch vehicles would both shorten development time and reduce costs when they entered serial production. On November 27, 1954, ARDC issued System Requirement No. 5 for a reconnaissance satellite. General Operational Requirement No. 80, issued on March 16, 1955, formally established an Air Force requirement for a reconnaissance satellite. The "ultimate" requirement called for a resolution that would "detect objects no more than 20' [feet] on a side." On April 2, 1956, WDD, which had acquired responsibility for the WS-117L from ARDC Detachment 1 at Wright-Patterson AFB, completed its development plan. It called for full operational capability in 1963, and a research and development cost of about $115 million. The WS-117L was to undertake more than reconnaissance; another version of the satellite would carry infrared equipment to detect Soviet missile launches. The Air Force contracted with RCA, Lockheed, and the Glenn L. Martin Company for design studies of a military satellite that would become operational in 1965. On June 30, 1956, the Air Force notified Lockheed that it had won the WS-117L contract. The military satellite was to use an Atlas booster with a Lockheed-built upper stage, later called Agena. Once in orbit, the Lockheed booster-satellite would be stabilized with its nose pointed toward the Earth, to allow a camera system to take pictures. The development plan was approved in July 1956 by the Air Staff. However, the Defense Department provided only $4.7 million for FY 1956. Even the conservative estimates of program costs indicated a need for at least $39.1 million in start-up funding.

Meanwhile, in a March 1956 report to the Air Staff, RAND recommended substituting film recovery in place of electronic scanning and readout in orbit. On June 26, 1956, RAND issued "Physical Recovery of Satellite Payloads: A Preliminary Investigation." The RAND proposals received support in late summer when a Defense Department study panel concluded that a solution to the reentry problem, then being investigated for ballistic missile warheads, was within reach.

Forging a National Space Policy

The status of the WS-117L reconnaissance satellite remained in doubt. In November 1956, Secretary of the Air Force Donald Quarles directed WDD to cease all efforts toward vehicle construction at Lockheed, and imposed sharp funding limits. An emerging civilian space

Original Air Force WS-117L Program Team in Inglewood, California, shortly after arrival at the Western Development Division, March 4, 1956, in front of Ramo-Wooldridge Building on Arbor Vitae, Inglewood, California. **Standing left-to-right**: Capt William O. Troetschel; Edwin Kolb, WPAFB; 1Lt John C. Herther; Lt Col William G. King, former program director at WPAFB; Russell Johnson, AFRADC; James Suttie, AFRADC; Joseph Fallik, AFRADC; Capt James S. Coolbaugh; Capt Frank S. Jasen. **Kneeling left-to-right**: Fritz Runge; Capt Richard P. Berry, AFCRC; Capt Robert C. Truax, USN, program director and chief engineer; Robert Copeland, WPAFB; Lt Col George P. Jones, AFCRC; Lt Col George Harlan, SAC liaison.

effort also affected the military WS-117L program. Back in 1955, when the WS-117L reconnaissance satellite program was beginning, American civilian scientists proposed that scientific satellites be launched as part of the upcoming International Geophysical Year (IGY), planned for late 1957-1958. The Technological Capabilities Panel also had urged the U.S. to launch a small scientific satellite to establish the precedent "freedom of space," allowing them and later military satellites to orbit over any country without prior permission. The intended precedent turned on the maritime legal principle "freedom of the seas," where ships of all nations have free passage on the high seas outside territorial waters. In the mid-1950s, no legal regime existed for outer space, and Administration officials believed establishing free passage with a civilian satellite would ensure the legality of passage for subsequent military WS-117L satellites. With such a precedent, the Soviet Union would be discouraged from attempts to destroy an overflying satellite.

In March 1955, the various civilian satellite proposals were sent by design to Donald Quarles, at that time Assistant Secretary of Defense for Research and Development. Convinced of the importance of freedom of space to the future of U.S. intelligence, he had privately encouraged the U.S. National Committee for the IGY to formally propose a scientific satellite. Its proposal was sent to the National Academy of Sciences, then back to Quarles for Defense Department review. On May 20, he submitted a proposal to launch an IGY satellite, and the national policy to guide it and the emerging reconnaissance satellite program, to the National Security Council (NSC). The satellite's peaceful purpose was to be stressed (it was to be a civilian-directed scientific activity), but the underlying intent was to establish the principle of freedom of space and the right of satellite overflight in international law. The NSC agreed on May 26, 1955, and Eisenhower approved the project and the policy the following day. Eisenhower announced the IGY satellite on July 29 (a few days after his return from the Geneva Summit Conference).

A scientific review panel selected the Navy's Vanguard satellite proposal which used an all-new booster developed from the Viking sounding rocket. Despite the importance of the principle it was to establish, the Eisenhower Administration kept its space policy a secret and held Vanguard to tight funding limits, to avoid interfering with high-priority ballistic missile programs. Program costs still increased from $20 million to $100 million by mid-1957. To avoid an international debate over military space activity that might jeopardize the principle freedom of space, government and military officials were prohibited from discussing any plans for such activities. The civilian Vanguard was to be first into space, while the WS-117L marked time.

In the summer of 1957, CIA Director Allen Dullas told Quarles, now Secretary of the Air Force, that the Soviet Union "probably is capable of launching a satellite in 1957." Despite this warning, Administration officials made no effort to speed up Vanguard or "race" the Soviets to launch the first satellite. The WS-117L also remained a low-priority, long-term effort. Even the ICBM program was subject to overtime restrictions and reductions in production and deployment. Secretary of Defense Charles E. Wilson showed open hostility to satellites, saying at one point that he "wouldn't care" if the Soviet Union launched a satellite before the U.S. On Friday, October 4, 1957, the Soviet Union obliged him by placing Sputnik 1 into orbit. The event had a stunning political effect on public opinion in the United States. An aide to Senator Lyndon Johnson (D-Tex) captured this mood:

> It is unpleasant to feel that there is something floating around in the air which the Russians can put up and we can't . . . . It really doesn't matter whether the satellite has any military value. The important thing is that the Russians have left the Earth and the race for the control of the universe has started.

In the following days and weeks, the Eisenhower Administration was judged complacent in the face of an impressive Soviet achievement. Politicians and pundits used the Sputnik launch to challenge everything from United States education, defense and space policy, to the size of automobile tailfins.

Soviet Premier Nikita Khrushchev quickly exploited this loss of confidence in the United States by ordering the launch of a second and larger satellite, Sputnik 2, which orbited the Earth on November 3 and carried a dog named Laika. This was followed on May 15, 1958 by Sputnik 3, a 1,327 kg geophysical satellite. The weights of these satellites were far beyond the lifting capabilities of U.S. rockets, sparked American concern over the size of the Soviet ICBM force, and prompted a "Missile Gap" controversy at the end of the decade. Air Force intelligence believed the Soviets had begun to deploy a very large ICBM force which would leave the U.S. vulnerable to a surprise attack in the near future. The Navy and Army estimated a small Soviet ICBM force would be deployed. CIA intelligence estimates fell in the middle range—somewhere between those of the military services. None of these estimates was based on hard information, as the few U–2 overflights had failed to disclose a single military ICBM site. Too small an area had been covered to prove there were no ICBM sites, however. The Missile Gap controversy affected domestic politics. Inflated estimates of a large Soviet missile force, leaked to the press, were used to attack Eisenhower defense policy. Democratic politicians depicted Eisenhower as "doing nothing" in the face of a Soviet missile build-up, intent on balancing the budget while he marked time playing golf. Deeply frustrated and unable to reassure the American public, President Eisenhower—given the U–2 information available to him— remained convinced that no missile gap existed. Yet, because the U–2 remained a secret, he refused to divulge the source that prompted his conviction.

In the "space race," as the press liked to term astronautical events, the launch of Sputnik 1 established the very legal precedent that Eisenhower sought. Soviet officials never asked permission for their satellite to pass over the United States. Only four days after Sputnik 1 was launched, Eisenhower and Quarles discussed the issue. Quarles observed, "the Russians have . . . done us a good turn, unintentionally, in establishing the concept of freedom of international space . . . . The President then looked ahead and asked about a reconnaissance [satellite]."

Organizing America's Space Effort

With subsequent U.S. space activities now inevitable, Eisenhower's first impulse was to assign all space activities to the Department of Defense. In early February 1958, he created the Advanced Research Projects Agency (ARPA) that was to oversee initial defense research and development, initially including both civil and military space activities. However, Vice President Richard Nixon and the chairman of the newly established President's Science Advisory Committee (PSAC), James R. Killian, convinced Eisenhower that a civilian national space agency would be preferable.

Eisenhower decided to use the National Advisory Committee for Aeronautics as the basis for the new space agency, and he submitted the proposed legislation to Congress on April 2, 1958. Signed into law on July 29, the National Aeronautics and Space Act created the National Aeronautics and Space Administration (NASA), which began operations on October 1, 1958. NASA inherited existing scientific satellites and planetary missions from the National Science Foundation and ARPA. Soon, NASA would add manned space flight to its responsibilities. Eisenhower split U.S. space activities between NASA (civilian) and ARPA (military). The military services now reported to ARPA on space matters, which significantly reduced the Air Force's role.

NSC directives expanded the basic outlines of U.S. space policy in June and August 1958 and December 1959. The first directive called for establishing a "political framework which will place the uses of U.S. reconnaissance satellites in a political and psychological context most favorable to the United States." The second directive said:

> Reconnaissance satellites are of critical importance to
> U.S. national security . . . . Reconnaissance satellites
> would also have a high potential use as a means of imple-
> menting the open skies' proposal or policing a system of
> international armaments control.

The third described the military space missions which were considered to be peaceful uses of outer space. At the same time, the 1959 directive noted, the United Nations Ad Hoc Committee on the Peaceful Uses of Outer Space now accepted the "permissibility of the launching and flight of space vehicles . . . regardless of what territory they passed over during the course of their flight through outer space." While freedom of space had been tentatively established in international law, the U.N. committee added this only applied to "peaceful" missions.

The Administration clearly counted reconnaissance satellites in the "peaceful" category of defense support missions. But it limited to studies any offensive military space operations, such as orbital nuclear weapons, satellite interceptors, space-based anti-ballistic missile systems, and lunar-based ballistic missiles lest they provide the Soviets with an excuse to attack reconnaissance satellites. A few days after Sputnik 1 was launched, Quarles directed the Air Force not to consider nuclear weapons in orbit in its future space planning. This extended to even the suggestion of an offensive mission. On October 20, 1958, ARPA Director Roy Johnson ordered the Air Force to stop using the "Weapons System" designation (such as WS-117L) for military satellites, "to minimize the aggressive international implications of overflight . . . It is desired to emphasize the defensive, surprise-prevention aspects of the system. This change . . . should reduce the effectiveness of possible diplomatic protest against peacetime employment."

The Eisenhower space policy can best be understood as strategic, rather than tactical, in nature. It was intended to establish the basic legal principles and missions for the long term. In this, it proved very successful. Eisenhower's space policy was less successful on the tactical level of the day-to-day Cold War political struggle with the Soviet Union. The Eisenhower Administration stressed that the U.S. was not in a "race" with the Soviets, dismissing such efforts as "stunts." American space flight activity was directed toward pure scientific research, civil applications (such as communication satellites), and limited military support applications (such as reconnaissance satellites). But this policy ignored the powerful symbolism that space held publicly in the late 1950s; space *was* part of the Cold War. The Soviets' aggressive drive in space contrasted strikingly with Eisenhower's apparent passivity. The American public looked to Eisenhower for *visible* leadership in space. By refusing to acknowledge the political importance of space activities, Eisenhower gave to the Soviets a major diplomatic advantage which fed the crisis atmosphere and the Missile Gap controversy. Ironically, attempts by Eisenhower and his scientist-advisers to downplay space activities conducted for purposes of international prestige resulted in the very things they wanted to avoid—demands for increased defense spending and a more aggressive space effort.

The Air Force WS-117L program, renamed Sentry and still later SAMOS (for Satellite and Missile Observation System), continued under development at the Lockheed facility at Sunnyvale, California. On January 22, 1958, the NSC assigned the highest national priority to development of an operational reconnaissance satellite. On February 3, Eisenhower ordered the highest and equal national priority for all ballistic missiles—the Atlas and Titan ICBMs and Thor and Jupiter IRBMs—and the WS-117L reconnaissance satellite and Ballistic Missile Early Warning System (BMEWS) radar network. A few days later he approved development of another reconnaissance satellite, code named Project CORONA, using Thor boosters and film recovery, as proposed by RAND two years before. This program, like the U–2, was placed under CIA control with the Air Force a joint venture partner.

Project CORONA

CORONA began with an assessment of overflight reconnaissance systems conducted by the President's Board of Consultants on Foreign Intelligence. On October 24, 1957, it issued a report that concluded the Air Force WS-117L project, for a variety of reasons, would not become operational until the early 1960s. With the launch of Sputnik 1 and the emerging Missile Gap controversy, board members recommended that an interim photo reconnaissance satellite system could be developed and flown more quickly. This system would fill the gap between the time the U–2 ceased to be of service and the time the WS-117L became operational. Deputy Secretary of Defense Donald Quarles and CIA Director Allen Dulles endorsed the proposal.

Unlike WS-117L, which by now had been identified in the press as a reconnaissance satellite, this new program was to be managed covertly. For the first several months, in fact, no written records were kept of meetings and briefings. Eisenhower's final approval of a revised project using the Itek 24-inch focal length camera in April 1958, it is said, was handwritten on the back of an envelope. The President approved Richard Bissell, CIA Special Assistant for Planning and Development who previously guided the U–2 project, as the head of Project CORONA. His Air Force counterpart once again would be Brig. Gen. Osmond J. Ritland, now Vice Commander of the Air Force Ballistic Missile Division.

Original plans called for CORONA to consist of an extremely simple spin-stabilized satellite, similar to the RAND proposals, that used a reentry capsule to return the film and camera to Earth. The Itek Corporation in early 1958 proposed using a camera patterned after those of the WS-461L reconnaissance balloon program. This system returned only the film, but promised photographs of three times better resolution than those of the original camera design, but it required that the satellite be stabilized in space with the camera pointed towards the Earth. While spin-stabilization was proven, attitude stabilization on all three-axes represented a greater technological challenge. Bissell agonized before finally deciding to abandon the spin-stabilized satellite design in favor of the Itek system.

The Lockheed Missiles and Space Division was to integrate the various hardware elements, and to accomplish this effort it received a one-page work statement. CORONA would use a Lockheed Agena upper stage booster-satellite containing the Itek camera (later called the KH-1 [for "Keyhole"]), and a General Electric reentry capsule that would return the exposed film to Earth.

Because the legal question of freedom of space remained open, Project CORONA was named "Discoverer" to establish a cover. It was described as a scientific project that conducted biomedical research and other experiments in space. Discoverer 1 failed at launch on February 28, 1959. But this was only the first in a series of failures. Boosters exploded during launch, satellites tumbled in orbit, the film, exposed to the harsh conditions of space, turned brittle and broke, and attempts to separate the capsules from the Agena and secure reentry into the Earth's atmosphere at the proper time and place, seemingly proved futile. As one problem was solved, another one took its place. Despite a growing string of failures, Eisenhower's support of the Air Force-CIA CORONA team never wavered.

The National Reconnaissance Office

The United States lost its main source of intelligence on the Soviet military force structure and capabilities on May 1, 1960, when a U–2 was shot down deep inside the Soviet Union and its pilot, Francis Gary Powers, was captured. President Eisenhower publicly admitted he had authorized the mission, and, in the international crisis that followed, ordered an end to aerial overflights of the Soviet Union.

With U–2 intelligence gone and reconnaissance satellites not yet operational, Eisenhower asked his science advisers to conduct two reviews of U.S. intelligence operations. One looked at intelligence organizations, while the other evaluated the Air Force SAMOS program. Up to this point, the Air Force had planned and directed the SAMOS program, with the Strategic Air Command scheduled to control the operational satellite system. The U–2 had been run by the civilian-controlled CIA supported by the Air Force. Presidential Science Adviser George B. Kistiakowsky conducted the SAMOS study with the assistance of two PSAC staff members. On August 18, 1960, Kistiakowsky met with Air Force Undersecretary Joseph V. Charyk, Carl F.G. Overhage, Director of MIT's Lincoln Laboratory, and Edwin Land to discuss the findings. SAMOS, they agreed, had both technical and management problems (the first launch had yet to take place), and that direction should be removed from the Air Force and transferred to a new civilian office in the Defense Department. This office would control program management, policy, plans, priorities, and orbital operations.

Administration officials judged strategic reconnaissance from space, like the U–2 aerial overflights, to be a national intelligence resource critical to the safety of the United States. It should not be controlled by any single military service or even a single intelligence agency such as the

CIA. On August 22 Kistiakowsky briefed Richard Bissell, the CIA's CORONA director, on the recommendations. The following day he briefed Defense Secretary Gates on the plan, which Gates endorsed. Meanwhile, as this reconnaissance satellite reorganization was being hammered out, CORONA achieved success. A test capsule from Discoverer XIII containing an American flag was successfully recovered after reentry on August 12, 1960. The capsule from Discoverer XIV containing the first images of the Earth taken from space was recovered a few days later, on August 18, 1960.

At 8:15 a.m. on August 24, 1960, shortly before the NSC meeting was to start, James Killian, Edwin Land, George Kistiakowsky, and National Security Advisor Gordon Gray met with President Eisenhower. They briefed him on the first satellite reconnaissance photos returned from space only a few days before. To avoid provoking the Soviets, Eisenhower decided that no reconnaissance satellite photos ever should be made public. Subsequently, NSC members discussed Kistiakowsky's recommendation for a new Air Force office to direct and control the SAMOS program; both Secretary Gates and President Eisenhower gave their approval. Air Force Secretary Dudley C. Sharp issued the necessary directives on August 31, 1960 that established the "Office of Missile and Satellite Systems." Air Force Undersecretary Joseph Charyk became the first director of the new office, and in this capacity reported to the Secretary of Defense.

To oversee the analysis of the satellite photos, President Eisenhower also acted on the recommendations of another intelligence review. The photo interpretation sections of the Air Force, Navy, and Army were combined with the CIA's Photographic Intelligence Division to form a single, civilian-controlled group. The National Photographic Interpretation Center (NPIC), which would report to the Director of Central Intelligence, held responsibility for the processing, analysis, and distribution of reconnaissance photos (both airborne and satellite). Arthur Lundahl, who had formally headed the CIA division, was named NPIC Director. When John F. Kennedy became President in January 1961, he endorsed Eisenhower's organization of space activities, now divided into civil, military, and reconnaissance missions. The new Defense Secretary, Robert S. McNamara, asked Charyk to remain as Air Force Undersecretary and as the first Director of the renamed National Reconnaissance Office (NRO), the existence of which remained a state secret. Three decades would pass before the NRO would be officially acknowledged. Under the Kennedy Administration the words "SAMOS" and "Discoverer" disappeared from public statements and the U.S. government no longer acknowledged satellites were used for reconnaissance—a policy that remained in effect until 1978.

CORONA operations continued behind a veil of secrecy. Originally intended as a short-term, interim effort, CORONA's intelligence achievements became so important that it was extended as an operational project. During 1960 and 1961 the CORONA satellites carried a series of improved cameras—the KH-2 and KH-3—which provided higher resolution of images and improved reliability. A major breakthrough came in February 1962 with the first launch of a KH-4 camera, which allowed stereo photos to be taken from orbit. Later CORONA satellites carried two reentry film capsules. The final camera was the KH-4B, introduced in September 1967. While the film from the original KH-1 on Discoverer XIV had a resolution of 35 to 40 feet, the KH-4B had a resolution of 5 to 6 feet. Moreover, satellite lifetime on orbit increased from one day, with Discoverer XIV, to between 18 and 20 days for the later CORONA satellites. Technical problems became the exception rather than the rule. The 145th and final CORONA satellite was launched on May 25, 1972. Its second capsule, recovered on May 31, brought the program to a close.

The establishment of the NRO ended direct Air Force control of satellite reconnaissance. After 1960, Air Force space activities focused on the launching and tracking of satellites, and conducting the previously defined military support missions: communications, missile early warning, meteorology, navigation, and the detection of nuclear detonations on Earth and in space.

II

Manned Military Space Flight

Although the Eisenhower Administration had organized U.S. civil and military space activities, one issue remained: what role would the Air Force play in U.S. manned space flight? NASA's manned space flight activities, approved by the President, derived from its charter to pursue science and civil applications in space. The Air Force space role, on the other hand, now was limited to the identified and approved defense support missions performed by unmanned, automated satellites. How might manned space flight serve these missions?

The Air Force manned space flight effort grew out of earlier aeronautical research. To understand the aerodynamic unknowns of supersonic flight, the Army Air Forces after World War II had built an experimental rocket-powered aircraft, the Bell X-1, designed to reach speeds above Mach 1 and survive. On October 14, 1947, Capt. Charles E. Yeager piloted the aircraft to fly faster than the speed of sound. Advanced versions of the X-1 reached speeds in excess of Mach 2 and altitudes of 90,000 feet. Another of the "X-planes," the Bell X-2 achieved a speed of Mach 3 and an altitude of 126,907 feet. At the same time, development work was underway on the North American X-15. This aircraft was designed to reach speeds of Mach 6 and altitudes above 60 miles. Aeronautical engineers in the 1950s thought that space would be reached in a series of such evolutionary steps, flying faster and higher until finally an experimental aircraft went into orbit. Such a vehicle might also serve as a replacement for existing and planned manned bombers. Starting in the early 1950s, Bell Aircraft conducted studies under the code names BOMI, Brass Bell, ROBO (Rocket Bomber), and HYWARDS. On October 10, 1957, ARDC Headquarters consolidated the Brass Bell, ROBO, and HYWARDS studies into a single project for experimental manned space flight called "Dyna-Soar," a contraction of "Dynamic Soaring."

The X-20 Dyna-Soar

The Dyna-Soar envisioned a three-step development program. Step I would develop an experimental hypersonic test aircraft. Step II would use a two-stage rocket booster to accelerate the aerospace vehicle and its single pilot to a speed of 18,000 feet per second and an altitude of 170,000 feet. It would then glide for 5,000 nautical miles, and could undertake reconnaissance or bombardment missions. Step III would extend the vehicle's performance to orbital flight, where it would be able to carry out reconnaissance or bombing missions.

On January 24, 1958, a few months before NASA's creation, the Air Force Deputy Chief of Staff for Development requested approval for an ambitious Air Force space program judged "essential to the maintenance of our national position and prestige": (1) Ballistic [missile] tests and related systems; (2) The X-15 and advanced research for manned space flight; (3) Dyna-Soar and its space reconnaissance and global bombardment subsystems; (4) WS-117L, a multi-satellite system for 24-hour reconnaissance, a manned military space station, and a strategic communications station; and finally (5) a manned military Lunar Base. Needless to say, much of this proposed military space program ran counter to the Eisenhower Administration's stated and unstated policies of minimizing the cost of space ventures, of avoiding wherever possible "prestige" missions, and, most importantly, of rejecting any kind of offensive military space activities that might place American reconnaissance satellites at risk. The Dyna-Soar component of this space program soon faced both political and technical problems.

In early November 1958, W. E. Lamar, Air Force Deputy for Research Vehicles and Advanced Systems, and Lt. Col. R. M. Herrington, Jr., the chief of the Dyna-Soar Program Office, briefed officials of both ARDC and U.S. Air Force Headquarters on program funding. General S. E. Anderson, the ARDC commander, stated that although he supported the Dyna-Soar program, he thought any references to space operations should be deleted before they were proposed to the Air Staff. Air Force officials decided to stress the suborbital aspects and the possibility of a military prototype system. When the Dyna-Soar presentation was given to Richard C. Horner, the Air Force Assistant Secretary for Research and Development, he bluntly warned the officers that if a strong space weapons system program was offered to Defense Department officials, Dyna-Soar probably would be terminated. On January 7, 1959, Deputy Defense Secretary Quarles approved $10 million for Dyna-Soar, but only to conduct a research and development program—and not, he stressed, as any recognition of Dyna-Soar as a space weapons system.

A few months later, on April 13, 1959, Herbert F. York, Director of Defense Research and Engineering, established Dyna-Soar's objectives. He specified its primary goal to be non-orbital flight up to speeds of 22,000 feet per second. Dyna-Soar would be launched by an existing

Artist's concept of a Dyna-Soar launching.

booster, maneuver during reentry, and make a controlled runway landing. Its secondary objectives were testing of military systems and orbital flight. York said these secondary objectives should only be implemented if they exerted no adverse effects on the primary objective. The Air Force source selection board subsequently chose the contractors for this Dyna-Soar program on November 9, 1959. Boeing would build the one-man aerospace glider, while the Glenn L. Martin Company would furnish a modified Titan I ICBM booster. Before development could begin, Joseph Charyk, who succeeded Horner as Assistant Secretary of the Air Force for Research and Development, ordered a "Phase Alpha" study. This reexamined a range of different configurations to solve the problems of manned winged reentry. When completed at the end of March 1960, Dyna-Soar emerged from the study unchanged and the Defense Department released development funding.

The Air Force completed its plan for the Dyna-Soar test flights in April 1960. The plan envisioned a total of 20 air-drop tests from a modified B–52, to begin in July 1963. Starting in November 1963, five unmanned suborbital launches were to be made from Cape Canaveral, followed in November 1964 by the first of 11 manned sub–orbital flights. These suborbital flights divided Step I into a three-part development effort. The first part would gain data on orbital flight and test military systems, including high-resolution photographic, radar, and infrared sensors, advanced bombing, navigation, and guidance systems, space-to-surface missiles, and rendezvous systems. The second part envisioned an interim military system able to conduct reconnaissance and satellite inspection, while part three was to be the fully operational weapon system. All three parts, limited to studies, remained only an outline.

Originally a modified Titan I ICBM was selected to boost the Dyna-Soar to suborbital speeds. By November 1959, this booster seemed barely adequate for Step I. Moreover, it could not be modified to lift the weights anticipated to orbital speeds. The Dyna-Soar Program Office proposed that a Titan II ICBM be substituted for suborbital flights, while a Titan II with a modified Centaur third stage be developed for orbital missions. The cost increase and delays incurred in the change would be minor. Air Force headquarters, however, did not approve the switch to a Titan II booster until over a year later, on January 12, 1961.

The Air Force had long wanted to undertake orbital flights with Dyna-Soar, and in October 1960 Brigadier General M. B. Adams, Deputy Director of Systems Development at Air Force Headquarters, instructed ARDC to develop a "stand-by" plan to achieve orbital flight with the Step I glider as soon as possible. In December 1960, the Dyna-Soar Program Office proposed using the same booster for both the suborbital and orbital flights; the program could be accelerated significantly. Although ARDC headquarters did not approve the "stand-by" plan, it set in motion major changes in the Dyna-Soar program.

On May 4, 1961, Boeing proposed a "Streamlined Program." This plan envisioned the elimination of the suborbital flights, the use of available subsystems, and switching to a Saturn C–1 as the booster. On August 3 and 4, 1961, Dyna-Soar program director Col. Walter L. Moore, Jr. briefed the Strategic Air Panel, the Systems Review Board, and the Vice Chief of Staff on the streamlined effort. By eliminating suborbital flights, the first air drop could be made in mid-1963, the first unmanned orbital flight in 1964, and the first manned flight in early 1965. This compared with January 1967 for the first manned orbital flight. Although the Air Force backed this proposal, exactly which booster would be used remained an open question. The three leading contenders were the Phoenix A388 (a new booster design using a solid propellant first stage and a liquid propellant second stage), the Soltan, a Titan II with two solid fuel 100-inch strap-on boosters, or NASA's Saturn C–1 booster. Heavier payloads such as Dyna-Soar required a booster that could bridge the gap between the Atlas Agena (at that time the largest Air Force booster) and NASA's 1-million-pound-thrust Saturn C–1. A modified version of the Soltan concept, called the Titan IIIC, was finally selected as the Dyna-Soar booster. This was a Titan II core with two 120-inch solid rocket boosters and a Transtage (a small third stage).

Air Force headquarters approved the Streamlined Dyna-Soar Program on December 11, 1961, tieing it to the Titan IIIC development program. The new plan envisioned 20 air drops beginning in January 1965 and completed in October 1965. The first unmanned orbital flight would be made in November 1965 (on the fourth Titan IIIC test launch), with the second in February 1966. The first manned launch would be made in May 1966, with the eighth Titan IIIC flight. All launches would be made from Cape Canaveral with the landings made at Edwards AFB. The first flights would be single-orbit missions; a multi-orbit mission would take place in November 1967.

On February 23, 1962, after a review, Defense Secretary McNamara directed that Dyna-Soar be reoriented—away from an operational system and toward an experimental program like those of the earlier X-planes. He also ordered the military name "Dyna-Soar" replaced with a numerical designation to reflect its research role. On June 26, 1962, Defense Department announced "X-20" as the new designation (although Dyna-Soar continued to be used by personnel within and outside the program). With the redesignation, the X-20 program had reached its final form. A winged, manned, recoverable spacecraft did not possess as large a payload as a manned capsule-type spacecraft (such as Mercury or Gemini) when launched on a given booster (an important consideration given the "lift-gap" with the Soviets). A winged spacecraft also took longer to develop and cost more than did a manned capsule. The Dyna-Soar program, always troubled by funding limitations and, more seriously, by the absence of a clearly defined military mission, now was judged by some as too small for such operational space missions as rendezvous and space station resupply. In September 1962, Secretary McNamara voiced these very concerns. Following another briefing on the program, he questioned Air Force officials whether Dyna-Soar represented the best expenditure of national resources.

Secretary McNamara was interested in the military potential of NASA's Gemini two-man spacecraft. When McNamara reoriented the X-20 program towards a research role, he suggested Defense Department participation in the Gemini program to demonstrate manned rendezvous. At a November 1962 meeting with NASA Administrator James E. Webb and Associate Administrator Robert C. Seamans, Jr., McNamara made a surprise proposal that NASA's Gemini be merged with planned Air Force Gemini flights, and that the combined effort be moved to the Department of Defense. NASA rejected the proposal on the grounds that a military program would interfere with Apollo and posed political and foreign policy concerns. Air Force leaders also did not like the proposal, which threatened the X-20. The results of the few "Blue Gemini" flights, they believed, would not be worth the cost.

McNamara nonetheless remained interested in the subject. On January 18 and 19, 1963, he directed a comparison study of the X-20 and Gemini to see which was the more feasible approach to military space flight. A few days later, an agreement between Defense Department and NASA allowed military experiments to be flown on board NASA Gemini missions. On March 15, he requested a further comparison be made between the X-20 and Gemini with respect to four prospective military missions—satellite inspection, satellite defense, reconnaissance, and orbiting of offensive space weapons. He added that a manned space station serviced by a ferry vehicle could be the most feasible approach. The Air Force review committee, meanwhile, completed its study on May 10, 1963, and found neither the Gemini capsule nor the X-20 able to meet any of the four missions without modifications. With use of a mission module and a Titan III booster, the Gemini would provide better orbital maneuverability and payload capacity. The X-20, however, had better reentry flexibility.

On July 22, 1963, Vice President Lyndon B. Johnson asked McNamara for his views of the space station's importance to military spacefaring. McNamara replied a few days later; he considered multi-manned, long-duration orbital flights most important to military space activities. Because there was no clearly identified manned military space mission, however, he believed that efforts should be directed towards providing experience in manned space flight. The Air Force participation in the Gemini flights would provide much of this experience. McNamara thought that a space station would be useful in conducting experiments on every type of military space mission, and that it might evolve into an operational system.

In August 1963, Harold Brown, who succeeded York as Director of

Defense Research and Engineering, authorized the Air Force to make a study of a manned military space station. The study, to be finished by early 1964, was to focus on the reconnaissance mission with the goal of assessing the military usefulness of man in space. To define the characteristics of the station, the Air Force was instructed to consider the use of such programs as the X-15, X-20, Mercury, Gemini, and Apollo. A few weeks later, on October 23, members of the X-20 project briefed McNamara and other Defense Department officials on the status and plans for that effort. Afterward, McNamara asked project personnel what the Air Force planned for the X-20 after maneuverable reentry had been demonstrated. He stated categorically that he could not justify spending $1 billion for any program in want of an ultimate goal. Until he understood what manned military space missions might reasonably be conducted, he was no longer interested in further X-20 spending.

The next day, Harold Brown reportedly offered a manned orbiting laboratory program to the Air Force in exchange for the cancellation of the X-20 Dyna-Soar. General Curtis E. LeMay, Air Force Chief of Staff, rejected the offer and ordered a rebuttal to the proposal be prepared. On November 14, 1963, Brown recommended to McNamara that the X-20 be cancelled in favor of a military space station program. He proposed two possible configurations. The larger design used the Apollo Lunar Module adapter, launched by a Saturn IB, as the station. The ferry vehicle would be an Apollo Command Module orbited by a Titan IIIC. Brown preferred the second option—a four-man military station which used the Gemini as ferry vehicle. Both the station and Gemini spacecraft would be launched with Titan IIICs. Management of the Gemini program would be transferred from NASA to the Defense Department by October 1965, with the first ferry launch in mid-1966, and space station tests underway in mid-1967. In the long term, a Lifting Body ferry vehicle could be developed to allow runway landings.

The new military space station plan ran into problems with NASA officials. They had no objections to a manned military space program, but they could not support a military space station. Instead, NASA officials suggested that the Air Force consider a manned laboratory that did not involve docking, crew transfer, or resupply. On November 30, 1963, Harold Brown made a new proposal that took into account the NASA objections. A Gemini and an attached 1,500-cubic-foot laboratory module would be launched by a Titan IIIC. The laboratory could be operated in space for 30 days and then abandoned, its two-man crew returning to Earth. This could develop into the Gemini-based proposal, which Brown still viewed as the most feasible design.

Despite last-minute efforts to keep it alive and link it with the emerging military space station program, the X-20's fate was sealed. On December 9, 1963, McNamara met with President Johnson, and the President agreed to cancel the X-20 Dyna-Soar. The following day, on

December 10, McNamara announced cancellation of the X-20; its tests of reentry technology would be undertaken by the ASSET program, a series of small rocket-boosted models. Dyna-Soar would be replaced by a Manned Orbiting Laboratory.

The Manned Orbiting Laboratory

The start of the Manned Orbiting Laboratory (MOL) program underlined the changing relationship between the military and the Defense Department. MOL had been suggested by Harold Brown, approved by McNamara, and its design influenced by NASA. The Air Force had no active participation in the deliberations or recommendations. Starting a program of this magnitude without directly involving the service that would direct it was unprecedented. Before work could begin, the MOL program had to survive a Pre-Project Definition Phase (Pre-Phase I). This was followed by the Project Definition Phase (Phase I). Not until Phase II would contracts be issued, hardware built, and launches made. And, as might be expected, early work concentrated on selecting the experiments MOL would conduct. Approval to begin development depended on producing an experiment package that McNamara and his advisers might accept as militarily useful.

The MOL program's primary objective was to qualitatively and quantitatively test the military usefulness of man in space. The experiments selection ground rules focused on the military role of man in space, covered the full spectrum of military applications, and made maximum use of ground tests and existing equipment to keep costs as low as possible. Testing man's role for space reconnaissance was acceptable as long as it did not produce an actual reconnaissance capability. That is, the initial objective was *not* to develop an operational manned military space system. Rather, MOL was to determine the role of man in space and how his unique capabilities could be used in various military space activities. MOL presumably would show if sufficient justification existed to develop a manned military capability in the future.

The baseline MOL design was a two-man Gemini B spacecraft attached to a Laboratory Vehicle which would be launched as a single unit by a Titan IIIC. MOL's role as a test program was reflected in the launch and experiment plan. The spacecraft would be launched from Cape Kennedy into a 125-250 nautical mile equatorial orbit, inclined less than 36 degrees to the equator. This orbit would not pass over any part of the Soviet Union and it was not envisioned that "reconnaissance quality" photos would be taken. Once in orbit, the two-man crew would not perform any extra-vehicular activity, but transfer to the Laboratory Vehicle through a hatch cut in the Gemini B's heatshield. The crew would then spend 30 days in space operating the experiments. At the end of the mission, the crew would transfer back to the Gemini B, reactivate its systems, re-enter the atmosphere and splash down in the ocean.

Artist's concept of the Manned Orbiting Laboratory on its Vandenberg launch pad. After the cancellation of the MOL program, the SLC-6 pad was modified for polar orbit launches of the military Space Shuttle. Following the loss of *Challenger*, the Air Force "mothballed" SLC-6.

Despite the studies, MOL development stalled. Although plans were submitted, the experiments were judged insufficient and their costs too high. Some officials proposed a combined Air Force/NASA space station program. With unmanned automated military spacecraft now performing successfully on orbit for extended periods, some in the Department of Defense argued that man had no military role in space. Other studies examined different MOL configurations which combined Apollo hardware and a one-man Gemini capsule. Finally, various medical experts declared that humans could not survive 30 days in a weightless orbital environment. MOL program goals consequently shifted during 1964 and early 1965, and moved towards operational-type missions. The launch site changed from Patrick AFB, Florida, to Vandenberg AFB, California, where the MOL could be launched into a polar orbit allowing overflight of the Soviet Union; manned space reconnaissance became its primary mission.

Not until August 25, 1965, a full 20 months after the program began, did President Johnson announce that development work would start. The MOL configuration remained the original baseline design. The contractors: McDonnell Aircraft would build the Gemini B reentry capsules, Douglas Aircraft would furnish the Orbital Laboratory, and General Electric would fabricate the experiments. Officially, MOL's reconnaissance mission remained unannounced. The first MOL test launch, scheduled for November 1966 from Patrick AFB, would be an unmanned suborbital test of the heatshield hatch. Subsequently two unmanned launches of the complete MOL vehicle would take place from Vandenberg AFB, the first scheduled on April 15, 1969. The first of five manned MOL launches was scheduled on December 15, 1969. Although no follow-on operational

Artist's concept of a MOL-Titan IIIC launch. The solid- propellant boosters have separated and are falling away. The two-man crew in the Gemini spacecraft atop the stack, would transfer to the orbiting laboratory module (beneath it) for a 30-day mission.

24

system was authorized, the Air Force officers connected with MOL felt confident that such a program would eventually be approved. With President Johnson's announcement in August 1965, it seemed that the Air Force now had a role in American manned space flight.

The MOL program, like other space projects, suffered from technical, weight, and budget problems. By February 1966, technical difficulties had caused a nine-to-twelve-month slip in the schedule. A weight increase in the MOL vehicle required that the Titan IIIC launch vehicle be replaced by the Titan IIIM, a rocket which used seven-segment strap-on boosters (rather than the five-segment strap-on boosters of the Titan IIIC). Despite these early difficulties, a Gemini heatshield test was successfully completed on November 3, 1966. The launch vehicle, a Titan IIIC, carried an unmanned Gemini with the hatch cut into its heatshield, and a Titan II tank which simulated the laboratory. The Gemini, released during the ascent, made a suborbital splashdown near Ascension Island in the Atlantic. The hatch came through the fires of reentry intact. The Titan IIIC continued to ascend and placed the simulated laboratory into orbit.

Far more serious were funding constraints. In 1965-1966, Congress moved to pay for new social programs to create President Johnson's "Great Society." And an increasing U.S. involvement in Vietnam and its related military demands on resources affected MOL from the start of its development. Initial funding limits in 1966 could be met with relatively minor technical and schedule changes, but the real "crunch" came in the fall of 1967. The Air Force had estimated that $600 million was needed for initial MOL development. The FY 1968 budget included only $430 million. The Air Force estimated that this would result in a 15-month delay in schedule, but the Defense Department permitted only a one-year delay. The first unmanned launch would now take place in late 1970, with the first manned mission scheduled in the summer of 1971.

As technical problems multiplied, the funding situation continued to deteriorate. Estimates of MOL funding for FY 1969 ranged from $600 to $640 million. In June 1968, Defense Department officials advised the Air Force that no more than $515 million would be provided. That resulted in program deletions and elimination of some tests as well as changes in assembly requirements and contractor hardware tests. The first manned MOL flight was pushed back another three months to the fall of 1971. MOL's funding problems intensified after the election of President Richard M. Nixon at the end of 1968. About $700 million for each of the next three years was seen as necessary to keep MOL a viable project. Only $500 million seemed the maximum that would be made available. In early 1969, another round of budget cuts occurred. Not even major changes in the development plan could now accommodate them. The fifth manned MOL flight was cancelled and the first unmanned flight was delayed to early 1971, while the first of the four manned MOL missions slipped from the fall of 1971 to mid-1972.

Four years had passed since the MOL program began. Under the original plan, the first manned launch was to have been made in 1969. In 1969, however, the first launch remained three years in the future. The original cost of the program, estimated at $1.5 billion, by 1969 had doubled to $3 billion. Finally, within the Air Force and the Defense Department, the question of identifying a significant role for military men in space was debated again. In the aftermath of the Apollo fire at Cape Kennedy that claimed the lives of three astronauts, all hands recognized that ensuring the safety of astronauts predominated—above and beyond any role for men in orbit that might be identified. Underscoring that perception at an Air Force meeting devoted to the subject, one official snapped: "His [the astronaut's] most important role will be tending to the life support system."

On June 10, 1969, Defense Secretary Melvin R. Laird informed Congress that MOL would be cancelled. By way of explanation, he cited the need to reduce military spending and recent advances in the performance and reliability of automated military satellites. In fact, President Nixon cancelled MOL to permit work to begin on another automated reconnaissance satellite project. A total of $1.3 billion had been spent on the program. The estimated savings from this cancellation over the next three years was $1.5 billion. The end of MOL brought to a close Air Force attempts to create a separate military manned space program. This outcome, however, would spare the service the sharp competition for funds and division of interests between proponents of manned and unmanned space flight that in the years ahead would sorely afflict the nation's civil space agency, NASA.

Military Space Shuttle Plans and Operations

NASA's manned Space Shuttle program, intended to replace all existing U.S. boosters, began at almost the same moment that MOL expired. If a fleet of Space Shuttles were to serve as the universal replacements, this "man-rated," reusable launch vehicle would have to be economical to fly and maintain, and meet all existing and future launch requirements, such as resupplying NASA's planned space station and the needs of the Air Force and Department of Defense—including launches into polar orbit, a large payload bay, and the ability to make cross-range maneuvers during reentry. To begin the venture, in 1970 the NASA Administrator Thomas Paine cancelled production of the Saturn launch vehicles, the nation's only true heavy-lift launchers. Another launch vehicle, preferably the Space Shuttle, now had to replace them.

Much as NASA influenced the design of MOL, Air Force desires had a major impact on the Shuttle design. Working with NASA, in January 1973 the Air Force formed the Defense Department Shuttle User Committee to coordinate military Shuttle use and identify potential military applications. Operations from Vandenberg AFB were planned to start in December 1982. The User Committee also developed plans for phasing-out "obsolete expendable boosters," such as the Atlas, Delta, and Titan. Air Force involvement with the Shuttle program, nonetheless, would be far different than its association with the Dyna-Soar and MOL programs. Unlike these separate military space programs directed and controlled by the USAF, military operations for the Shuttle would be limited to launching satellites and conducting experiments on orbit.

To support military Shuttle operations, the Air Force authorized a new control center in 1979. The Consolidated Space Operations Center (CSOC) would consist of two elements. The Satellite Operations Center (SOC) would handle on-orbit control of instrumented Defense Department satellites, while the Shuttle Operations and Planning Center (SOPC) would be used for planning and control of manned military Shuttle missions. The CSOC would be built at Falcon AFB, near Colorado Springs, Colorado. Until it was in operation, the Air Force control center at Sunnyvale and NASA's Johnson Spaceflight Center would be used.

The Shuttle also allowed non-astronauts to go into space. Payload Specialists would operate experiments and be involved in the deployment of satellites. They might include scientists, contractor personnel, and, for Air Force and Defense Department Shuttle missions, military officers. To supply these military Payload Specialists, the service inaugurated the Manned Spaceflight Engineer (MSE) program at Los Angeles AFB in August 1979. The MSE program trained qualified military officers (both Air Force and Navy) in the design, development, and integration of satellites and experiments to be flown on the Shuttle. The MSEs would develop plans and operational procedures to be used by NASA astronauts on military Shuttle missions. These plans included crew activities, payload training for NASA astronauts, Shuttle-satellite integration, and coordinating medical and scientific experiments sponsored by military agencies. During military Shuttle flights, they would also act as Payload Communicators for the ground control at Sunnyvale and Falcon AFB. As Payload Specialists, the MSEs underwent Shuttle and zero-G familiarization and participated in NASA simulator training and operational planning. In all, 32 MSEs (29 men and three women) were selected and trained between 1981 and 1986. Unlike the Dyna-Soar and MOL astronauts, however, the names of the MSEs were not made public for some time.

Air Force use of the man-rated Shuttle remained a point of controversy. One problem was manned space flight history—the Air Force had put a tremendous amount of effort and money into the Dyna-Soar and MOL programs, yet had nothing to show for it. The other was control—NASA ran the Shuttle program; the Air Force, on the other hand, purchased and for the most part controlled disposition of the nation's expendable boosters. As might be expected, technical problems, inadequate funding, and politics caused years of delay for the Shuttle. After launches

Launch of the STS-51C mission on January 24, 1985. This was the first military Space Shuttle flight that carried a classified payload.

began in April 1981, the Shuttle experienced continuing problems and launch delays. The start of launch operations at Vandenberg AFB was repeatedly delayed by difficulties of constructing the SLC–6 Shuttle pad. Finally, as events turned out, Congress paid for a fleet of only four Shuttles. Just how these four would meet all of America's space launch needs, civil and military, remained an open question. Air Force officials debated continuing to buy expendable boosters even though the Shuttle was now operational. Undersecretary of the Air Force Hans Mark supported employing the Shuttles to launch all military satellites, while most senior Air Force officers opposed a policy that depended exclusively on the Shuttle. The Martin Marietta Corporation had developed the Titan 34D as a replacement for the Titan IIIC and D boosters, pending full replacement by the Shuttle.

While the debate swirled, in 1983 Air Force Vice Chief of Staff General Jerome F. O'Malley reopened the issue of expendable boosters. And on December 23, 1983, Air Force Undersecretary Edward Aldridge, Jr. issued a memo, "Assured Access to Space," which called for the production of improved Titan 34D boosters to launch military satellites as a back-up to the Shuttle, with only two launches per year made. The remaining military Shuttle flights would amount to about twenty percent of the total number of Shuttle missions each year. In early 1985, Martin Marietta's Titan 34D-7 (later Titan IV) was selected. And a medium-lift

The crew of STS-51C. **In the front row, left to right**: Loren J. Shriver (Pilot) and Thomas K. Mattingly II. **In the back row are**: Gary E. Payton (Payload Specialist), James F. Buchi and Ellison L. Onizuka (Mission Specialists). Payton was the first Manned Spaceflight Engineer to make a Shuttle flight. Onizuka was killed in the Challenger explosion a year later.

launch vehicle was needed, since the Atlas, Titan II, and Thor production lines had been shut down to ensure a full Shuttle launch schedule. In February 1984, Aldridge ordered a study of alternative launch options for Shuttle military payloads. Study participants found the Global Positioning System (GPS) Block II navigation satellites as the most urgent military space project requiring expendable launch vehicles.

The first military Shuttle mission, STS-51C, carried a classified payload into orbit on January 24, 1985. On board was the first MSE to fly, Major Gary Payton. The Shuttle returned to Earth on January 27. This was followed on October 3, 1985 by the second military Shuttle flight—STS-51J (which was also the first flight of Space Shuttle Atlantis). Major William A. Pailes served as the MSE for this mission. When completed on October 7, the mission became the second and last time an MSE would travel into space. The year 1986 was to have been an ambitious one for the Shuttle program, including a first launch into polar orbit from Vandenberg AFB. But on January 28, 1986, Space Shuttle Challenger exploded 73 seconds after lift-off, killing its seven-member crew. Two Titan 34D failures and the loss of a Delta booster during the same time period grounded the whole U.S. space program—NASA, commercial, and military.

Following the loss of Challenger, the Air Force terminated plans to use Vandenberg AFB for Shuttle launches. SLC–6 was mothballed and its equipment transferred to NASA, other Air Force projects, or the Navy.

Defense Support Program early warning satellite "Liberty" as it was deployed from the STS-44 mission. The DSP satellites provided warning of Iraqi Scud missile launches during the Gulf War, which allowed civilians to take cover before the missiles impacted and Patriot missile batteries to engage the incoming Scuds.

The Shuttle Operations Planning Center also was cancelled in 1987, owing to a change in Shuttle payload policy. President Ronald Reagan announced that Shuttle no longer would be used for all American satellite launches. In future, it would carry only research payloads or payloads specifically requiring the presence of men in space. And because the backlog of military payloads awaiting launch, caused by the Shuttle grounding, military Shuttle flights only would continue for a short while longer. Because Payload Specialists would not be flown on board the early missions once Shuttle flights resumed, the MSE program also was terminated.

Expendable launch vehicles returned to favor. The purchase of more heavier-lift Titan IVs was authorized, and, on May 9, 1986, Air Force Headquarters ordered procurement of an improved Medium Launch Vehicle. On January 21, 1987, McDonnell Douglas' Delta II proposal was selected. In addition, surplus Titan II ICBMs were modified for use as space boosters. Finally, General Dynamics began development of advanced versions of the Atlas Centaur—the Atlas I and II. It would be the late 1980s and early 1990s, however, before they would be ready to launch payloads into orbit. In the meantime, NASA's Shuttle flights resumed with STS-26 on September 29, 1988. The next flight, STS-27 (December 2-6, 1988) was the first post-Challenger military Shuttle flight, followed by STS-28 (August 8-13, 1989). Although the payloads on these flights remained classified, the STS-33 mission (November 22-27, 1989) did have several unusual features. Up to this point, military Shuttle flights had crews who were all serving military officers. STS-33's crew included F. Story Musgrave and Kathryn C. Thornton, the first and only civilians to fly on a military Shuttle mission. (Thornton was also the only woman.) The mission commander was Colonel Frederick D. Gregory, the first African-American to command a space mission. STS-36 (February 28-March 4, 1990) and STS-38 (November 15-20, 1990) were the two military flights for 1990.

The final two military Shuttle missions marked breaks with the tight security of earlier flights. STS-44 (November 24-December 1, 1991) was the first and only satellite deployed by a military Shuttle flight to be unclassified. The payload was a Defense Support Program (DSP) Early Warning satellite dubbed "Liberty." The crew included Gregory and Musgrave from STS-33. Also on board was Army Chief Warrant Officer Tom Hennen, the first military Payload Specialist to fly since Payton and Pailes in 1985. Hennen was not a member of the MSE program, however. (He was also the first enlisted man to fly into space.) Hennen operated a manned reconnaissance experiment called Terra Scout. STS-53 marked the ninth and final military Shuttle flight (December 2-9, 1992). Although details of the satellite remained classified, the blackout only lasted through payload deployment. Once this was completed, normal communications resumed. With the landing of STS-53, the Air Force's involvement with the Space Shuttle and military manned space flight ended.

III

Military Space Operations 1958-1991

The Air Force emerged from the organization of the U.S. space program between 1958 and 1960 with the defense support missions of missile early warning and space defense. The first mission employed automated satellites equipped with infrared sensors to detect Soviet missile launches; the second embraced development of a satellite inspector vehicle, satellites to detect nuclear explosions on Earth and in space, and an Earth-based system of radars and cameras that could track satellites in orbit. At the same time, the Eisenhower Administration assigned several other military space application missions to other services and agencies. The Navy was to develop a navigation satellite system, while the Army would oversee strategic and tactical communications satellites. NASA acquired responsibility for developing meteorological satellites to serve both civil and military users.

Missile Early Warning - MIDAS to DSP

In 1956, the CIA warned the National Security Council Planning Board that "it is possible the USSR, if it sought full strategic surprise, could launch an attack on the continental U.S. without undertaking any observable preparations which would provide strategic warning." Although the probability of such an event remained low, the U.S. could not absolutely count on advance indications and warning of a Soviet surprise attack. Thus, considerable effort was devoted to building a radar network first to detect incoming Soviet bombers before they reached their targets, and, second, with the emergence of the Soviet long-range missile program in the mid-1950s, on larger Ballistic Missile Early Warning System (BMEWS) radars that could provide 10 to 20 minutes warning. Even with this radar warning, however, only "alert" aircraft with crews at the ready would have time to take off; most U.S. bombers would be caught on the ground in a Soviet missile attack.

In 1955-1956, Joseph Knopow, a Lockheed engineer, and Sidney Passman and William W. Kellogg of RAND began looking at ways to detect missile launches using infrared sensors. At first they considered equipping aircraft flying outside Soviet airspace with sensors. Again, due to the curvature of the Earth, aircraft could not observe most of the boost phase. But satellites also could carry the detection equipment. From a vantage point in space, infrared sensors on board satellites could detect the hot exhaust of a missile as it rose above the atmosphere. This would add five to eight minutes to the radar warning time, allowing more bombers to take off before Soviet warheads struck. Passman and Kellogg wrote in RAND Research Memorandum RM-1572:

During the early stages of the (ICBM) take off there is more than enough infrared emissions, but the Earth gets in the way . . . . After burnout there is not nearly enough infrared signal to give detection at any useful range.

The figures . . . lead one to speculate on the increased warning time and perhaps more accurate trajectory prediction that might be possible by getting around this geometrical limitation with a very-high-altitude search station—perhaps with a satellite-borne infrared search set.

The proposals for satellite detection attracted the attention of several scientific advisory committees, and a missile early warning system was incorporated into the WS-117L at the outset, in 1956. In February 1958 when ARPA assumed control of WS-117L, money was unavailable to start development. ARPA also envisioned a technical development program, rather than a crash effort to achieve a missile detection system at the earliest possible date. In September 1958, the Air Force Ballistic Missile Division which oversaw WS-117L, proposed an expedited development plan to Air Force Headquarters and ARPA. In late 1959 the Defense Department removed control of all military satellite programs from ARPA and returned direction of them to the Air Force. This included the early warning satellite program, which was to be made operational as soon as possible. (This was the time of the Missile Gap controversy.) The program was originally "Subsystem G" in the WS-117L program, then identified as the "Attack Alarm System" (AAS), and finally to be termed "Missile Detection and Alarm System," or MIDAS.

The MIDAS used the Lockheed Agena booster-satellite, stabilized on orbit in a nose-down attitude, with an infrared scanner and telescope mounted in a rotating turret on the nose. The infrared sensor traced out a circular pattern on the Earth below. MIDAS would detect the launches and relay information on the number of missiles, their launch sites, and azimuths (direction of travel) to ground stations. Given existing technology, as conceived, MIDAS would employ eight satellites placed in 2,000-nautical-mile-high polar orbits, equally spaced in two orbital rings. This would insure at least one MIDAS satellite above the Soviet Union at all times. The data on missile launches would be transmitted to readout stations located in Alaska, Greenland, and England. These MIDAS satellites were to have an operating lifetime of one year.

But MIDAS 1 launched on February 26, 1960, was lost when its Atlas-Agena launch vehicle exploded. MIDAS 2 successfully orbited on May 24, 1960; however, after 16 orbits the telemetry system failed. In June 1960 two radiometer experiment missions, RM-1 and RM-2, were added to the program. They were subsequently flown in December 1960 and February 1961 on Discoverer satellites to measure the background radiation on Earth and in space. By the time these flights had been completed, doubts expressed about MIDAS were reaching the American public. During testimony in the spring of 1961, McNamara advised Congress of "a number of highly technical, highly complex problems associated with this system . . . . The problems have not been solved and we are not prepared to state when, if ever, it will be operational." The Air Force revised its development plan in March to include 24 to 27 launches, with operations pushed back to January 1964. The service launched MIDAS 3

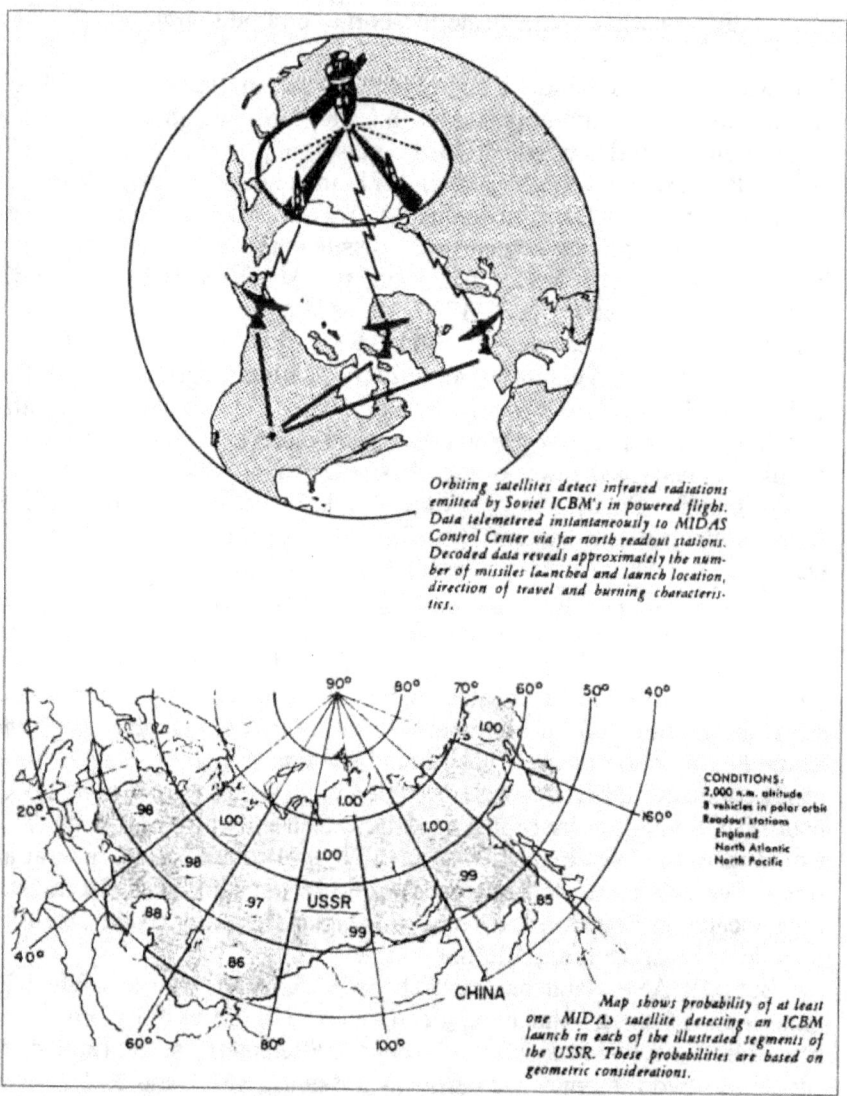

Orbiting satellites detect infrared radiations emitted by Soviet ICBM's in powered flight. Data telemetered instantaneously to MIDAS Control Center via far north readout stations. Decoded data reveals approximately the number of missiles launched and launch location, direction of travel and burning characteristics.

CONDITIONS:
2,000 n.m. altitude
8 vehicles in polar orbit
Readout stations
England
North Atlantic
North Pacific

Map shows probability of at least one MIDAS satellite detecting an ICBM launch in each of the illustrated segments of the USSR. These probabilities are based on geometric considerations.

Diagram showing an early plan for MIDAS operations, using a network of low altitude satellites to provide continuous coverage of the Soviet land mass.

on July 12, 1961 and it performed well for several orbits, but, before it could detect a missile launch, it suffered a power failure. MIDAS 4 followed on October 21; it reached orbit but could not be stabilized. Most of its systems operated for a week before a power failure occurred. Although press accounts claimed that MIDAS 4 detected a Titan II launch, this was erroneous.

Despite these setbacks, the Air Force pressed for a go-ahead for an operational system. In September of 1961, Harold Brown, Director of

Launch of a MIDAS satellite from Cape Canaveral. The MIDAS served as the test bed for the DSP satellites that followed.

Research and Engineering, organized an Ad Hoc group to review MIDAS. On completion of the review on November 30, 1961, the group identified three major technical problems associated with MIDAS. First was target detection—measurements of the Titan II (which was similar to the Soviet SS-7 ICBM) indicated that the MIDAS might have only marginal sensitivity against large liquid-propellant ICBMs. The group also speculated that MIDAS "is probably not effective" against smaller solid-propellant missiles such as Minuteman and Polaris, and expressed doubts that it ever could be made effective.

The second problem involved concerns that sunlight reflecting from clouds in the upper atmosphere might cause false alarms. The report said:

> As the MIDAS [infrared] equipment scans, highlights in the illuminated clouds of extent comparable to a missile plume will appear indistinguishable from a missile . . . . If the system threshold were low enough, the highlight would cause a false target . . . .
>
> Incomplete as these measurements are, data from balloons, U–2 flights, and Discoverer satellites (RM-1 and RM-2) give a more or less consistent picture which indicates that the false target problem will certainly be a severe one . . . .

The final difficulty involved system reliability. MIDAS was designed to operate on orbit for one year, a period much longer than any comparably complex spacecraft had achieved previously. Contractor estimates indicated a far shorter lifetime—between two weeks and a month. The report concluded, "The existing design of MIDAS is so complicated that it probably cannot be made reliable enough with the present and foreseen reliability art to warrant deployment." Although the Ad Hoc group's report considered MIDAS to have a valuable and needed role, questions arose over the requirement for satellite early warning in the post-1965 time period. McNamara directed the Secretary of the Air Force to submit an analysis on the value of early warning and MIDAS's role. Meanwhile, on July 31, 1962, the Defense Department reoriented MIDAS to a research and development effort. Its immediate objective was gathering radiometric background and target measurement data. In keeping with the secrecy that now attached to military satellites, the name "MIDAS" was replaced with "Program 461."

Two Program 461 launches took place during 1962. An April 9 launch was successful, but the spacecraft experienced a power failure shortly after attaining orbit, while another on December 19 ended in another failure and shower of debris into the ocean—this time the Pacific. Two more background measurement flights (RM-3 and RM-4) made during the year provided "irrefutable evidence" that eliminated the concerns over large target detection and background radiation clutter.

But the performance of Program 461 vehicles remained open to question. Testifying before Congress, Brown claimed that "the way the program was [currently] going, it would never produce a reliable, dependable system." The Defense Department cut FY 1963 funding to $75 million, while the FY 1964 budget was only $35 million. Because of the cuts, the Air Force dropped all deployment plans and directed its efforts toward improving reliability and detection probability, reducing false alarms, and attitude stabilization of the satellite. The long years of effort were finally rewarded on May 19, 1963. A 461 satellite launched on that date operated for six weeks and detected nine Titan II, Atlas E, Minuteman, and even Polaris missiles. The detection of solid-propellant Minuteman and Polaris launches was especially gratifying given the earlier doubts that MIDAS infrared sensors could ever spot these rockets. This success was followed by another launch on July 18. Although the satellite operated for only 96 orbits before starting to tumble, it successfully detected an Atlas E launch—as well as Soviet rocket tests. The twin successes showed that satellite early warning definitely was feasible and achievable.

On November 7, 1963, the Defense Department halted further Program 461 launches to allow development of an improved system able to detect submarine-launched missiles and IRBMs, as well as pinpoint their launch sites. This MIDAS follow-on program called for eleven Atlas Agena launches between January 1966 and July 1968, when limited operations would be achieved. This would be followed between March 1969 and August 1970 by launches of the operational system. These satellites would use a 6,500-nautical-mile-high orbit. This plan was later abandoned and the U.S. early warning satellite program was again reoriented. Now, a few MIDAS satellites would be placed in geosynchronous Earth orbit. At that altitude, three or four satellites could constantly watch the entire Earth; that made unnecessary a network of eight or more MIDAS/461 satellites in relatively low orbits. In 1967-1968, TRW won the contract to build the operational geosynchronous early warning satellites. Launches of these satellites began in the early 1970s, with the effort redesignated as the Defense Support Program (DSP). The 32.8 feet-long DSP satellite was launched by a Titan IIIC into a geosynchronous orbit.

The heart of the DSP satellite remained a large infrared telescope fixed to the front end of the spacecraft. The DSP pointed towards the Earth and the entire satellite was spun at a few revolutions per minute. This allowed the telescope to scan the Earth, watching for missile launches. The TRW design was elegantly simple, and had greater reliability, because it eliminated the rotating turret and slip rings used on Lockheed's MIDAS design. The infrared energy from a rocket was detected by sensors inside the telescope. As the sensors swept across an infrared source, an electronic signal was generated. The signals were relayed to a ground station and quickly passed to NORAD headquarters in Colorado. Once analyzed, the DSP data showed the number of missiles, their azimuth, and projected

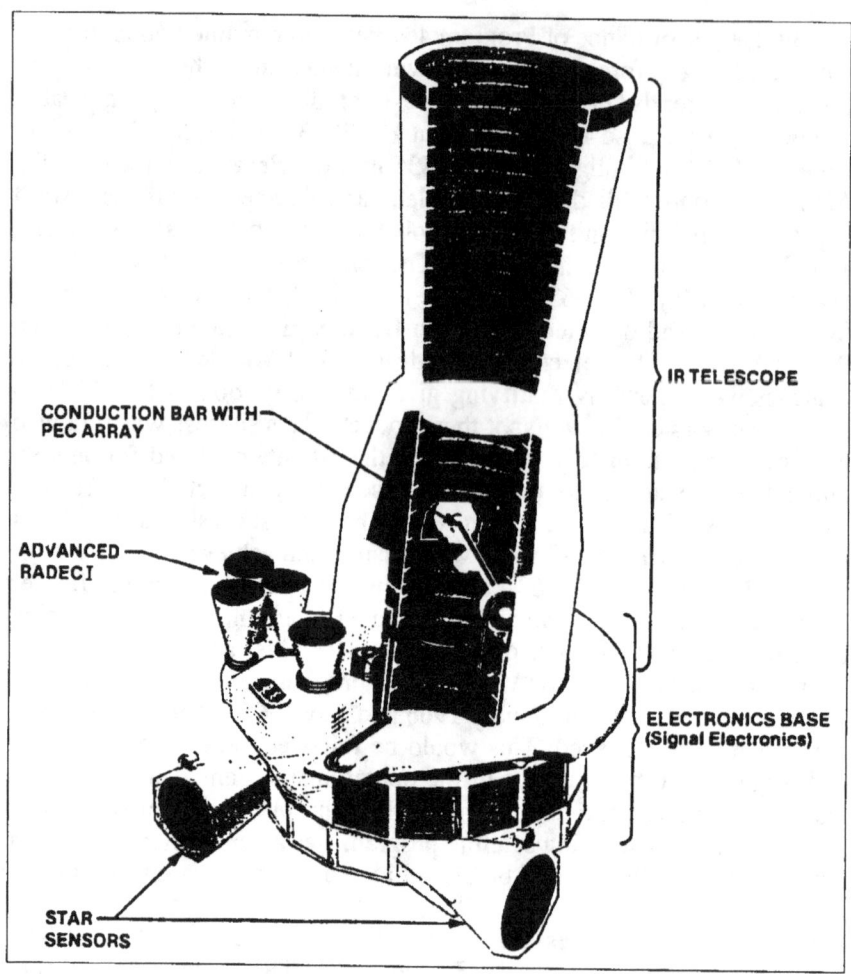

IR TELESCOPE

CONDUCTION BAR WITH PEC ARRAY

ADVANCED RADEC I

ELECTRONICS BASE
(Signal Electronics)

STAR SENSORS

Drawing of the DSP telescope system. Unlike the MIDAS satellite, which used slip rings to rotate the telescope, the DSP telescope was fixed to the spacecraft and used the satellite's rotation to scan. Because it was placed in a high altitude geosynchronous orbit, only a few DSPs were needed instead of eight-to-twelve MIDAS satellites in orbit at low altitudes.

impact points. The telescope was designed to detect large ICBMs and submarine-launched missiles. Events would show it had a far greater capability.

The DSP proved highly successful. In 1972, according to press reports, the DSP was declared operational and turned over to NORAD. Over the next two decades, DSP launches continued and improvements were made in the satellites themselves. In the event of a missile attack against the U.S., the DSP spacecraft would provide the President and the U.S. military a warning within moments of launch. By removing the possibility of a successful surprise attack, during the 1970s and 1980s the DSPs served to keep Soviet leaders from seriously contemplating such an action in the first place. In 1991 the early warning mission performed for the first time in a wartime role. U.S. and Coalition forces, as well as the civilian population of cities across the Middle East, relied on the DSP satellites for warning of Iraqi Scud missile attacks. In the wake of Desert Storm, General Donald J. Kutyna, Commander of U.S. Space Command in 1991, termed DSP the most important of all Air Force satellite projects.

The SPADATS/SSN Tracking Network

Another space defense mission involved Air Force responsibility for tracking all satellites and debris in orbit. This effort was composed of three elements—the Air Force Spacetrack system, the Navy's Space Surveillance System, and the Canadian Forces Air Defense Command Satellite Tracking Unit. Additional data also was provided by the Smithsonian Astrophysical Observatory. The complete network was called the Space Detection and Tracking System (SPADATS), operated by NORAD, located in Colorado.

The sensors that provided the raw data for SPADATS were scattered across the world. Spacetrack radars included the Cobra Dane located at Shemya Island, Alaska, the Milstone radar at Westford, Massachusetts, the AN/FPS-85 at Eglin AFB, Florida, and the AN/FPS-79 at Pirinclik (formerly Diyarbakir), Turkey. Additional radars at Vandenberg AFB, the Navy's Pacific Missile Range, and Cape Canaveral also contributed to SPADATS. The tracking radar on the island of Trinidad in the West Indies, for instance, was a major source of data. Although committed to a primary mission of early warning, the BMEWS radars at Thule, Greenland; Clear, Alaska; and Fylingdales Moor, England provided about one quarter of the total observations. The main Navy contribution, the Space Surveillance System, consisted of a series of ground stations stretching in a line across the southern United States that went into operation during 1959. Three transmitter sites sent a fan-shaped "fence" of radio signals thousands of miles into space. As a satellite passed through the fan, an echo was reflected back to receivers at the ground stations.

From 1960 into the early 1990s a network of Baker-Nunn cameras also provided tracking data based on optical sightings on a world-wide basis. These cameras could spot a satellite as small as a basketball out to 25,000 miles and were used to detect satellites orbiting beyond the range of other sensors. The Baker-Nunn cameras operated in twilight or darkness when the satellite was illuminated by the Sun but the Earth's surface was in darkness. When the satellite's position against a known star field was measured, very precise data was produced. The Air Force Baker-Nunn cameras sites were Sand Island (near Johnston Island) in the Pacific; Jupiter, Florida; Harestua, Norway; Santiago, Chile; Mt. John, New Zealand; San Vito, Italy; Pulmosan, South Korea, St. Margarets, New

Brunswick and Cold Lake Alberta in Canada, and Edwards AFB, California. Not all ten sites were in operation at the same time; most started operations between 1960 and 1977, while the phase-out of sites began in 1964. Another 12 Smithsonian Astrophysical Observatory Baker-Nunn cameras around the world also provided scientific data to SPA-DATS.

In the late 1980s and early 1990s, the NORAD tracking system, renamed Space Surveillance Network (SSN), underwent a major upgrade. The old Baker-Nunn cameras were replaced in the early 1980s by the Ground-based Electro-Optical Deep Space System (GEODSS). The original three sites, at Socorro, New Mexico; Choe Jong San, South Korea; and Maui, Hawaii, went into operation on March 1, 1983. A fourth site, on Diego Garcia Island in the Indian Ocean, started operation on January 15, 1987. The South Korean site closed in mid-1993. These cameras could provide real-time optical tracking data. The last two Baker-Nunn cameras, at San Vito, Italy and St. Margarets, New Brunswick, closed in 1991 and 1992. (The cameras are now used for scientific research, rather than Air Force tracking.) A spin-off of GEODSS was the Transportable Optical System (TOS). TOS was developed by MIT's Lincoln Laboratories and used a 21-inch telescope and electronic camera on a modified Nike-Ajax radar mount. The complete system could be crated and moved in a C–141 transport. Air Force Space Command tested the portable TOS at San Vito, Italy, between June and December 1991. (The Baker-Nunn camera closed down at the same time.)

Another recent improvement involved the Deep Space Tracking System (DSTS). Unlike radar, DSTS used passive receivers which detected the electronic emissions of a satellite. Direction finding and short baseline interferometry determined a satellite's position as a function of time. DSTS began as an Air Force Systems Command research and development program in 1988. The data would update the master catalog of objects in space, improve coverage in the eastern hemisphere, reduce the number of "lost" satellites, and monitor satellite maneuvers.

Air Force Space Command's network of radars was also expanded and improved. In the 1980s and 1990s the BMEWS radars were upgraded and three new early warning radars were added to the satellite tracking network. These were the Pave Paws West, located at Beale AFB, California, and Pave Paws East, at Otis AFB, Massachusetts, designed to provide early warning of submarine-launched missiles. The PARCS radar at Cavalier, North Dakota, originally built as part of the Army's Safeguard ABM system, was also used for satellite tracking. A long-standing gap in coverage was an inability to track some Soviet satellite launches on their first orbit. To close this gap, the Air Force established the "Pacific Barrier." Two of the radars, one at Kwajalein Island and the other at San Miguel in the Philippines, were in operation by the mid-1980s. A third radar, at Saipan, was under construction. The Pacific Barrier ran afoul of

Air Force budget cuts—the Saipan radar, though completed, never went into operation, while the San Miguel site was closed down. Orbital data obtained from this network were sent to NORAD's Space Defense Center inside Cheyenne Mountain, near Colorado Springs, Colorado, where it was processed by computers to produce the orbital predictions.

These data were used for several purposes. One was simply cataloging the ever-changing population of space vehicles and debris in various Earth orbits. Atmospheric drag causes satellites in low orbits to re-enter the atmosphere and incinerate. Any debris which survives may impact the Earth and cause damage or injuries. The location of a vehicle's reentry track and possible impact can be predicted, and, if necessary, warnings can be given. With the increasing population of space debris, the risk of collisions among debris and satellites also increased. The orbital data is used to warn when a man-made object will pass close to a Space Shuttle. Several times, Shuttles have had to maneuver to avoid such close passes. Of the more than 5,000 objects tracked in space, seventy-five percent was debris or fragments, twenty percent consisted of inactive payloads, while operational satellites amounted to only about five percent.

Nuclear Detection - Vela Hotel to IONDS

The automated Vela nuclear detection satellites remain one of the most successful Air Force space projects. The Air Force had been involved early in detection and monitoring of Soviet nuclear testing, beginning with Project Mogul in the late 1940s. This mission became more critical in the mid-to-late 1950s when the U.S., Soviet Union, and Great Britain began serious negotiations for a complete nuclear weapons test ban treaty. Any such agreement would require technical systems to police compliance with treaty terms and detect any violations. After conducting studies, on September 2, 1959, Defense Department directed ARPA to begin Project Vela. Named after the Spanish word for "watchman," it consisted of three elements. Vela Uniform employed sensors on Earth that improved the capability to monitor underground nuclear tests. Vela Sierra also involved Earth-based systems to detect atmospheric and space tests. Vela Hotel was a satellite system designed at first to scan above the horizon and detect nuclear tests in space.

Responsibility for Vela Hotel, as with other military satellite programs, transferred from ARPA to the Air Force on September 18, 1959. The service organized the project as a two-part effort. The first involved flying piggy-back proton, electron, neutron, X-ray and gamma ray detectors on board Discoverer satellites to provide background data on space radiation. The second would develop and launch actual Vela satellites. On November 24, 1961, the Air Force selected TRW as contractor to build ten Vela satellites. They were to be launched in pairs on five Atlas Agenas from Cape Canaveral in 1963-1964. The satellites, shaped as facetted spheres, carried X-ray, gamma ray, and neutron detectors which could

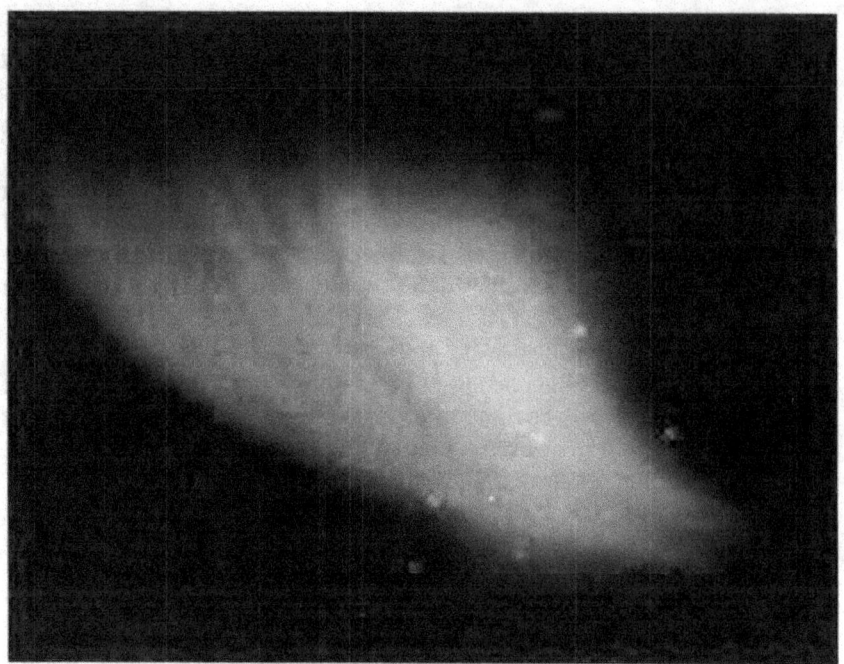

Photograph of the Kingfisher high-altitude nuclear test of November 1, 1962. The early Vela satellites were designed to detect such tests in space to police the Limited Test Ban Treaty. Later satellites were able to detect nuclear explosions in the atmosphere and outer space.

spot a nuclear explosion as far as 100 million miles from Earth. The Velas would go into 70,000-nautical-mile-high orbits, about a fifth of the way to the Moon, which placed the satellites above the Van Allen radiation belts. Each of the pairs of satellites would be placed on opposite sides of the Earth, thereby providing complete coverage of outer space.

While Vela was under development, negotiations continued on a complete nuclear test ban treaty. But the problems of detecting small underground nuclear tests with seismometers and disagreements over on-site inspections proved too great. By early 1963, efforts shifted to a more limited goal. On August 5, 1963, the U.S., Soviet Union, and Britain signed the Limited Nuclear Test Ban Treaty which prohibited testing nuclear devices in the atmosphere, underwater, and in outer space. Underground testing, though restricted, was still permitted. The treaty went into effect on October 10, 1963. Six days later, the Air Force launched the first pair of Vela satellites. These satellites proved highly successful, and on July 10, 1964, Defense Department announced the number of launches would be reduced from five to three, for a savings of $26 million. A more advanced design would be developed for the later missions. The second pair of Vela satellites was launched on July 17, 1964, and the third and

An early Vela satellite readied for launch.

final pair on July 20, 1965. The satellites had a design lifetime of six months, but they exceeded this by a factor of ten—the first pair operated until August 1968, the second until December 1969, and the final two were shut down in July and October 1970.

In March 1965, TRW was awarded a contract for a series of advanced Velas. Although they had the same general design, they were heavier and, fitted with improved optical detectors which could spot the flash of a nuclear fireball within the Earth's atmosphere. (France and Communist China had not signed the Limited Test Ban Treaty and continued to make atmospheric tests.) The equipment for detection of space tests was also improved. This raised the satellite's weight to 509 pounds, compared with the 334 pounds of the original Velas. (The last two Velas' weight would be increased further to nearly 800 pounds.) Because of this increased weight, the pair of satellites would be launched by a Titan IIIC.

Launch of the advanced Velas first occurred on April 28, 1967. This was followed on May 23, 1969 by a second pair, and the final pair on April 8, 1970. Although a total of 12 Vela satellites was launched, a maximum of eight were operational at any one time. For the final 18 years of operations, four satellites made up the network. The design lifetime of 18 months also was far exceeded—the final Vela operated for *14 years*. As the satellites got older, ground controllers had to exercise special care. Twenty-one years of Vela operations ended at 11:57 a.m. Pacific Daylight Time, September 27, 1984, when Col. Thomas Niquette, Director of Operations for the Air Force Technical Applications Center sent a command to shut down the final Vela satellite. (This was the first of the 1969 pair.) The twelve satellites had operated for a total of 108 years, traveled 3.23 billion miles, and detected about 40 atmospheric tests. They also had verified, contrary to the fears of some, that all State signatories to the nuclear test ban treaty had abided by its terms for over twenty years.

This did not mean the end of nuclear detection as an Air Force mission. Rather than building specialized nuclear detection satellites, Vela instruments now were carried aloft on other automated military satellites, such as the Defense Support Program (DSP) and Navstar GPS navigation satellites. The program was redesignated the Integrated Operational Nuclear Detection System (IONDS). The first Navstar to carry the IONDS sensors was the sixth Navstar, launched on April 26, 1980. This effort at detecting nuclear detonations has continued since then without interruption.

Communication Satellites - IDCSP to Milstar

Of all the civilian applications satellites which emerged in the 1960s, none affected everyday life more profoundly than did communication satellites. With them it would become possible to pick up a phone and call nearly anywhere in the world. It was also now possible for a news event to be covered live on television and relayed around the world via satellite. Science fiction writer Arthur C. Clarke first proposed radio and telephonic communications from satellites positioned in geosynchronous orbit. In the October 1945 issue of *Wireless World* magazine, he described three large satellites placed in 22,300-mile-high orbits over the equator. Rotating around the world at the same speed as the Earth's peripheral velocity, each would remain positioned above one point on the equator, making it possible to relay television signals from satellite-to-satellite around the world. (The concept was independently formulated in February 1947 by James Lipp in one of the early RAND studies.)

Although the emphasis in both satellite studies and development in the early and mid-1950s turned on scientific and reconnaissance applications, interest in communications applications continued. The first communication satellite was Project Score, an Atlas B placed into low Earth orbit on December 18, 1958. It carried a tape recorded Christmas message from President Dwight Eisenhower. In October 1958, ARPA directed the Army to begin work on the prototype of a communications relay (repeater) satellite called Courier. The first one was launched two years later on October 4, 1960. Operating as a "repeater" satellite, a message was transmitted to Courier, where it was rebroadcast it to a ground station.

The other communications satellite concept studied by both NASA and Defense Department in the late 1950s involved a "reflector" communication satellite. As the name suggests, the transmission was reflected from the satellite's surface back to the ground. A reflector satellite required no electric power or major on board systems. NASA tested the concept with two Echo balloon satellites. The military counterpart was Project West Ford. Rather than a balloon satellite like Echo, West Ford used 400 million copper dipoles, each 0.7 inches long and thinner than a human hair. The "needles" were carried in a canister on board a MIDAS. After release, plans called for the needles to slowly separate from the canister and form a ring around the Earth 2,000 miles high, five miles wide, and 25 miles thick. Transmissions could then be reflected off the ring. The first West Ford canister was launched on board MIDAS 4. Although it separated, the needles did not disperse. A second attempt was made on May 9, 1963 and was successful. Communication tests were conducted, and no harmful effects were noted on optical or radio astronomy from the ring, but, in the face of vigorous scientific protests, no subsequent dipole needle launches were made. Although Echo had shown reflector communication satellites were feasible, rapid advances in space and communications technology shifted attention to geosynchronous repeater satellites for both civil and military use.

In 1959, as a follow-on to Score, ARPA requested that the Air Force and Army prepare a joint development plan for a geosynchronous communication satellite. The booster and spacecraft were the responsibility of the Air Force, while the communications equipment, both on the ground and the payload on board the satellite, was to be overseen by the Army. When development began, there were three projects—Steer, Tackle, and Decree. The program soon ran into budget and technical criticisms, and in February 1960 the three projects were combined into a single major effort called "Advent." In September 1960, ARPA transferred Advent to the Army. The complex, 1,250-pound Advent payload was to be launched by an Atlas Agena, which would place test vehicles into 5,600-nautical-mile-high orbits. Subsequently, Atlas Centaur boosters would launch Advent satellites into geosynchronous orbits. A total of ten launches were planned, but work on Advent moved slowly. The division of authority between the Air Force and Army proved awkward and caused many problems. Cost estimates sky-rocketed—from $130 million in December 1958 to $352 million less than two years later. The Atlas Centaur booster, meanwhile, slipped two years behind schedule; all the while satellite technology advanced rapidly. A lighter-weight communication satellite could now perform Advent's mission.

On May 23, 1962, Defense Secretary McNamara cancelled Advent. As a replacement, the Institute for Defense Analysis recommended two systems. The first was a series of 20 to 30 small satellites in medium-altitude orbits. A number of these satellites could be orbited on a single booster. Later, a geosynchronous system would be developed. During the summer of 1962, McNamara accepted these recommendations. The Air Force was assigned responsibility for the launch vehicles, satellites, and the communication instruments they would carry, eliminating the split in responsibility that had caused problems with Advent. The Army would develop the ground terminals.

The medium-altitude satellites became known as the Interim (or Initial) Defense Communications Satellite Program (IDCSP). Development work was delayed for more than two years, however. The main problem now was the establishment of the Communications Satellite Corporation (COMSAT) in early 1963. McNamara thought that the Defense Department might lease COMSAT circuits to save time and provide COMSAT with an assured income, but the problems with such dual use soon became clear. Military and civilian transmission modes were completely different, requiring separate repeaters on each satellite. Civilian satellite links would serve large cities where the high traffic volume assured revenue, while military requirements were unpredictable and involved remote sites without adequate communications links. The satellite design requirements were also contradictory—COMSAT needed large numbers of low-power channels, while the military needed fewer, but more powerful channels. On July 15, 1964, McNamara dropped the idea of joint usage and authorized development of the IDCSP system.

The IDCSP satellites, three feet in diameter, weighed only 100 pounds and had no moving parts. Electrical power came from solar cells; there were no batteries. This meant the satellites would shut down whenever they were in the Earth's shadow. Philco-Ford won the prime contract for the IDCSP satellites, while Hughes Aircraft contracted to build the ground terminals. The IDCSPs could provide five commercial or eleven tactical quality two-way voice circuits, or 1,500 teletype circuits. With a design lifetime of three years, each satellite had a timer which would cut off power after six years, to prevent cluttering of the radio spectrum.

Originally, the IDCSP satellites were to be launched by Atlas Agena boosters into 5,000 to 6,000 mile-high orbits. By the fall of 1964, when the IDCSP satellites entered the final design and production phase, another possibility appeared. Titan IIIC test flights had been scheduled in the mid-1960s to carry sand or water ballast. John H. Rubel, Deputy Director of Defense Research and Engineering, thought this a wasted opportunity. He urged that the early Titan IIICs carry IDCSP satellites. This approach would accommodate up to eight satellites per launch and a higher altitude. Because their design was not easily adaptable to a geosynchronous orbit, the satellites were modified to use an 18,200-nautical-mile orbit. With an orbital period of about 22 hours and 20 minutes, the IDCSP satellites would drift slowly across the sky.

The first seven IDCSP satellites were successfully launched on June 16, 1966 on board the fourth Titan IIIC flight. Communication tests were conducted by the Army between Fort Dix, New Jersey, and Camp Roberts in California, before operations commenced. In the second IDCSP launch on August 26, all eight satellites were lost when the payload shroud failed 80 seconds after lift-off. A new shroud was designed and the third IDCSP launch took place on January 18, 1967. The eight satellites reached orbit successfully. The Titan IIIC launch of July 1, 1967 carried only three IDCSP satellites, a Despun Antenna Test Satellite, the LES 5 tactical communications test satellite, and the Defense Department SAGE 1 satellite which returned the first television pictures of Earth from geosynchronous orbit.

During July 1967, the Pacific IDCSP military communication network transferred from a research and development test phase to operational status. This provided strategic communication links among ground stations in Hawaii, the Philippines, South Vietnam, and the continental United States. A final launch on June 13, 1968 brought the total to 26 IDCSP satellites. The complete world-wide network was now considered operational, with the satellites subsequently renamed the Defense Satellite Communications System I (DSCS I). The IDCSP/DSCS I satellites demonstrated a remarkable lifetime. By late 1971, about 20 of the satellites were still operating (15 was considered adequate.) Although several shut down when their six-year lifetime was up, as late as mid-1976 three were still in use.

The IDCSP/DSCS I satellites were intended for strategic military communications between fixed bases. For tactical communications, however, the requirements were more severe. A portable ground terminal had to be small and lightweight enough to be carried by small ships, aircraft, or troops in the field. Under such conditions, the big dish antennas of conventional ground terminals were impractical. By using selected frequencies, it was possible to use a non-directional fixed antenna. However, because such an antenna could pick up both direct signals and reflections (such as from the ocean's surface), specialized receiving equipment was needed. The space technology for tactical military communications was first tested with the Lincoln Experimental Satellite (LES) program. Launched on December 21, 1965, on a Titan IIIC, an upper stage failure left LES 4 stranded in a 105-by-18,200-nautical-mile orbit. Despite this unplanned orbit, the satellite's beacon was used to analyze the problem of reflected transmissions. The LES 5, launched on July 1, 1967, along with three IDCSP satellites and two other test satellites, undertook communication tests with small mobile terminals on aircraft, ships, submarines, and jeeps. LES 6 followed on September 26, 1968.

In September 1965, a tri-service panel studying the future of tactical satellite communications recommended development of a large, high-power military geosynchronous satellite. The Defense Department awarded a contract for the satellite to Hughes Aircraft in December 1966. The Tactical Communications Satellite I (TACSAT I) was the largest such system yet built. Weighing 1,690 pounds, it was nine feet in diameter and stood 25 feet tall. The satellite could provide 40 voice or 700 teletype circuits to a terminal with a dish only three feet across. Because of funding limitations, however, only one qualification test article was built. After the ground tests, that satellite had to be refurbished for launch.

The Air Force launched TACSAT I on February 9, 1969, with a Titan IIIC and it went into a nearly perfect orbit over the Pacific. A month later, some 20 small terminals were linked together by the satellite. These included a portable transmitter weighing only 22 pounds and a receiver weighing only six pounds. The first year tested equipment, which limited it to civilian use—during the Moon landing in 1969, TACSAT I relayed live television to commercial stations in Alaska, providing for the first time in that state live coverage of a national/international event. On July 1, 1970, TACSAT I and LES 6 provided an initial operational capability for tactical military satellite communications. TACSAT I continued to operate for nearly four years. By this time, these early programs had set the stage for a fully operational satellite communication system.

The IDCSP/DSCS I satellites provided an initial operational military satellite communications system, but it was still far from what had been envisioned with Advent. In 1964, Defense Department studies of a geosynchronous military communication satellite system began in earnest. These continued through 1967, and in June 1968 the Defense Department approved start of the Defense Satellite Communications System II (DSCS II) program. The Air Force selected TRW as the prime contractor in March 1969. The firm was to build a qualification model and six flight satellites, which were to be launched in pairs on board Titan IIICs. These communication satellites would have both horn antennas for wide-area coverage and two steerable dish antennas for coverage of small areas. The first DSCS II pair, launched on November 2, 1971, experienced problems. Telemetry was received from the first satellite, but attempts to control it at first proved unsuccessful. No signals were detected from the second satellite, and it was lost in space. It took four days to locate the second satellite and transmit a new set of commands to both satellites, to bring them under control. In December 1971, both satellites began to rotate. The first satellite was successfully stabilized, but its antenna system was damaged in the attempt. Communications margins were reduced but it was still usable. After long analysis, the second DSCS II satellite was stabilized on June 8, 1972. Despite that success, it operated for only ten months.

Because of these problems, the DSCS II program underwent a major redesign. Not until December 13, 1973, did the Air Force launch the second pair of DSCS II. A third launch failed on May 20, 1975 owing to a launch vehicle guidance malfunction. The satellites went into a low orbit and re-entered the Earth's atmosphere and incinerated after six days. But with only one DSCS II satellite operational at this time, TRW was awarded a contract to build six more. The first pair of the new series was

Artist's concept of a Defense Satellite Communications System II spacecraft in orbit. The first DSCS II satellite, launched in 1971, became the first operational military communications satellite in geosynchronous orbit. The launch vehicle was the Titan III/34D booster. A total of 16 DSCS II satellites were launched.

successfully launched on May 12, 1977. The next launch, on March 25, 1978, was a failure. A second stage hydraulic failure prevented the Titan IIIC from reaching orbit. The Air Force successfully launched the final two of the second series of DSCS II satellites on December 13, 1978. This established the operational four-satellite network. An additional four DSCS II satellites were built—two were launched on November 20, 1979. The next DSCS II satellite, launched on October 30, 1982, was matched with the first of the DSCS III satellites. This was also the first launch of the Titan 34D booster, which replaced the Titan IIIC and D boosters.

Planning began for the DSCS III in 1973, while the DSCS II satellites were beginning to be launched. The new satellites carried multiple beam antennas to provide flexible coverage, counter jamming, and six active channels rather than the four of the DSCS II satellites. In 1977, General Electric won the Air Force contract to build one qualification and two

Artist's concept of the DSCS III satellite in orbit. This spacecraft was designed for greater flexibility with counter jamming, and it provided more channels that the DSCS II satellite.

demonstration satellites. During that same year, the House Appropriations Committee proposed leasing commercial circuits for military use (revisiting McNamara's proposal of 15 years before), and it eliminated funding for DSCS III. Nevertheless, the perceived problems involved in joint civil and military usage remained formidable, and funding was restored in a House/Senate committee. Following the October 30, 1982 launch of the first DSCS III satellite, deployment of the complete five-satellite network moved slowly. The Air Force launched the second and third DSCS III satellites in the mid-1980s, and the fourth in 1989. Not until the fifth launch in July 1993 was the full satellite "constellation" deemed operational.

The DSCS II and III satellites, like the original IDCSP satellites, were intended for strategic military communications between fixed bases. But tactical military satellite communications showed a similar growth during the 1970s and 1980s. The Navy's Fleet Satellite Communications System (FLTSATCOM) was a descendant of the 1960s LES satellites and TAC-SAT I. This program began in 1971, with TRW winning the contract a year later. The Navy funded the program and had responsibility for producing the surface terminals, while the Air Force oversaw development of the satellites, provided launch and tracking and control services, and used some of the FLTSATCOM's capacity. The Navy used one fleet broadcast channel (which provided 16 multiplexed teletype signals) and nine fleet relay channels. The Air Force portion claimed one wide-band and 12 narrow-band channels. The Air Force launched the first FLTSATCOM into a geosynchronous orbit on February 9, 1978. Four more followed by August 1981. Of these, one FLTSATCOM was damaged during launch and never became operational. Three more launches occurred between December

1986 and September 1989. One was lost during launch when its Atlas Centaur was struck by lightning.

The Ultrahigh-Frequency Follow-On (UFO) satellite replaced the FLTSATCOM series. Built for the Navy by Hughes Space and Communications, the Air Force launched UFO 1 on March 25, 1993. Regrettably, the Atlas I's first stage engine suffered an early shut down. Although the Centaur second stage continued to burn until the propellant was depleted, it could not reach the planned transfer orbit. UFO 2, launched on September 3, 1993, was successfully placed in geosynchronous orbit. The Navy has ordered a total of nine UFO satellites from Hughes.

Part of the Air Force use of the FLTSATCOM involved the Air Force Satellite Communications System (AFSATCOM), which relayed messages to and from nuclear forces such as bombers and missile silos, command posts and aircraft, and submarines. (Such a role had been planned for the original Steer program, which later became part of Advent.) The AFSATCOM transmissions are normally short, pre-formatted messages that required minimal power. The space elements of AFSATCOM originally consisted of transponders on board the FLTSATCOM and the Satellite Data System (SDS) spacecraft built by Hughes Aircraft. With the FLTSATCOM satellites in equatorial orbits and SDS spacecraft in polar orbits, AFSATCOM could provide world-wide coverage. The AFSATCOM reached initial operational capability on May 22, 1979 (shortly after the first FLTSATCOM was launched). At first, 15 B–52 bombers, four EC–135 command and control aircraft, and a ground station were equipped with receivers. Some 900 terminals eventually were to be deployed. The AFSATCOM was expanded with the launch of the DSCS III satellites. Each carried a single transponder for use by the system.

The follow-on to AFSATCOM and DSCS was the Milstar military communication satellite. Designed to use Extremely High Frequencies

Artist's concept of the Milstar communications satellite, designed to use Extremely High Frequencies (EHF) which do not suffer prolonged blackouts from high-altitude nuclear explosions.

(EHF) that do not suffer prolonged black-out from high-altitude nuclear explosions, the Milstar program began in October 1981, as part of President Ronald Reagan's upgrade of U.S. strategic forces. One element emphasized communication systems that could survive under conditions of nuclear war. Full-scale development began with award of the prime contract to Lockheed in July 1983. Tests were conducted using an EHF payload launched on board a FLTSATCOM satellite in December 1986. Development of these complex satellites took a decade; the Air Force launched the first of four Milstars on February 7, 1994. The vision of world-wide communications via spacecraft held by Arthur C. Clarke and his RAND contemporaries four decades before had been achieved.

Meteorological Satellites - DMSP

Development of weather satellites to serve all civil and military applications was originally assigned to NASA. A proposed military system was limited to studies while the Defense Department negotiated with NASA and the Commerce Department's Weather Bureau to develop a single joint civilian/military weather satellite system. NASA's Tiros I, based on an Army design and launched on April 1, 1960, began a revolution in weather forecasting. NASA's meteorological satellite program, however, could not meet military needs for coverage, readout locations, or timeliness. And NASA's entrant to meet all government meteorological requirements, a large satellite called Nimbus, was technically complex and long delayed in development (the first one would not be launched until the mid-1960s). Consequently, with approval from the Department of Defense, the Air Force in the early 1960s developed a separate, simplified, low altitude military weather satellite system to provide cloud cover photography. As more sophisticated instruments were developed, these would be added. (Indeed, this low-cost military meteorological satellite system, conceived, managed, and directed by Air Force Colonel Thomas O. Haig, proved so effective that the Weather Bureau soon advised NASA that it would buy near-copies of the military system instead of Nimbus.)

By the time these classified military weather satellites were in operation, U.S. involvement in the Vietnam War had grown into a massive commitment of U.S. forces, including air attacks against targets in North Vietnam. The Defense Meteorological Satellite Program (DMSP), as it was eventually called, became the primary short-term forecasting tool for tactical military operations, particularly in the air war. The DMSP satellites, placed in 450-nautical-mile polar orbits, passed over Southeast Asia at 7:00 a.m., noon, 7:00 p.m., and mid-night local time. The satellites carried equipment that provided both day and night visual imagery of cloud cover. (Civilian weather satellites provided daytime only photos.) The photos had two levels of resolution, 0.33 and 2 nautical miles, and covered an area 1,600 nautical miles across. The night images were used to find the differing cloud-top levels and breaks in the clouds. The infrared sensors also could spot burning rice fields, which allowed estimates of the smoke and haze coverage. The next day's target selection and mission plan were based on the DMSP photos from the 7:00 p.m. pass. The orders for the number and type of aircraft, as well as ordnance selection, used data from the mid-night pass. The day's air strikes were carried out based on the DMSP photos from the 7:00 a.m. pass. Air-to-air refueling tracks also had to be in areas that were free of clouds and turbulence at refueling altitude. Close air support and rescue operations also depended on the weather at low-altitudes.

In November 1966, the DMSP data produced 877 mission forecasts; of these 852 (97 percent) proved correct. On April 15, 1968, the Air Force informed the DMSP System Program Office that its satellite weather data had saved $25 to $28 million by eliminating the need for weather recon-naissance aircraft in Southeast Asia. Another study indicated that, at the height of the bombing effort, DMSP paid for its annual cost every three to six months. Two DMSP ground stations, located at Ton Son Nhut Air Base, South Vietnam, and at Udorn Air Base, Thailand, supported military operations with meteorological data. Each facility consisted of a pair of sandbagged vans and a large dish antenna. A separate van also was equipped to receive daytime photos from low altitude civil meteorological satellites. Unlike the encrypted DMSP photos, those from the Weather Bureau satellites could be picked up with simple receivers.

The Navy, with access to DMSP photos ashore, was relatively slow to make direct use of the DMSP at sea. Not until 1970 was an aircraft car-rier, the *USS Constellation*, equipped to receive DMSP images. The installation involved locating a temporary ground terminal on the hangar deck. The value of the DMSP photos was underscored by the Captain's willing-ness to give up an aircraft parking space (a quantity in short supply on any carrier) to house the terminal.

During the Vietnam War, distribution of the DMSP military meteoro-logical imagery was restricted and the satellites themselves classified. But in a May 1967 interview, General William Momyer, the commander of Air Force operations in Southeast Asia, alluded to their importance:

> As far as I am concerned, this [satellite] weather picture is probably the greatest innovation of the war. I depend on it in conjunction with the traditional forecast as a basic means of making my decisions as to whether to launch or not launch the strike. And it gives me a little bit better feel for what the actual weather conditions are. The satellite is something no commander has ever had before in a war.

Needless to say, with the cloud cover photos widely distributed in Southeast Asia, DMSP had become an open secret by the early 1970s. In 1973 Secretary of the Air Force John McLucas publicly announced the existence of the DMSP program, and DMSP photos were released for

Artist's concept of a Defense Meteorological Satellite Program model 5D orbiting the Earth. Since the DMSP went into operation during the Vietnam War, these satellites have provided weather data critical to combat operations.

civilian use. The most striking of the early photos was a night composite photo of the entire United States showing the lights of cities across the land. The DMSP's more advanced sensors expanded the capabilities of civilian forecasters. The data was supplied to the Weather Bureau's successor, the National Oceanic and Atmospheric Administration (NOAA). Much of the DMSP data is stored at NOAA's depository at the University of Wisconsin.

The DMSP program underwent major advances in the 1970s and 1980s. Block IVA and B satellites used in the 1960s were replaced in the early 1970s by the Block 5A, B, and C satellites. The Block 5C satellite was about five feet long, weighed 426 pounds, and was launched by the Thor-Burner II. By the final (unsuccessful) Block 5C launch on February 18, 1976, a total of 34 DMSP launches had been made over the previous decade. At the same time, a new generation of DMSP satellites was under development. The rectangularly-shaped Block 5D-1 satellite was four feet across and 19.3 feet long with its solar panel deployed on a boom. The Block 5D satellite weighed about 1,131 pounds (nearly three times that of the Block 5C), which required using an Atlas as the launch vehicle. The Block 5D-1 carried 300-pounds of instruments. Although a resolution of

0.3 and 1.5 nautical miles for the primary Operational Linescan System was only slightly improved over earlier DMSPs, the Block 5D-1 system provided this resolution over the entire photo. Other instruments included temperature/moisture sounders and aurora detectors. The infrared sounder provided plots of temperature and water vapor versus altitude within the atmosphere. The microwave temperature sounder measured temperature/altitude profiles within the atmosphere, even through clouds. The aurora detector measured the location and intensity of the aurora. Radar and long-range communication operators in the far north used this data to predict radio interference and service interruptions.

The Air Force launched the first DMSP Block 5D-1 on September 11, 1976. Although it experienced stabilization difficulties on orbit soon after launch, the satellite recovered and continued to operate for 36 months. The second Block 5D-1 launch occurred on June 5, 1977 and also ran into stability problems because the solar panel boom only partially deployed. The satellite stabilized on June 26, 1977. Subsequent Block 5D-1 launches, on April 30, 1978 and June 6, 1979 were successful, but the final Block 5D-1 launch, on July 14, 1980, failed. The follow-on satellite, the DMSP Block 5D-2, although similar in design, was both larger at 21 feet long and heavier at 1,656 pounds. The Block 5D-2 payload also increased to 400 pounds, and the Atlas F served as the launch vehicle. An additional instrument was a microwave imager. This passive microwave detector measured rain, soil moisture, sea state, and ice cover. The first Block 5D-2 was launched on December 20, 1980, followed by the second on November 17, 1983, with subsequent launches during the 1980s.

Just as the earlier DMSP satellites supported U.S. combat operations in Vietnam and Southeast Asia, later Block 5Ds supported tactical military operations in the 1980s. These included confrontations with Libya, as well as military action in Grenada in 1983 and in Panama in 1989. The DMSP data also figured in routine military operations, and alerted civil and military satellite operators to impending hazards from solar radiation. Such radiation showers can damage spacecraft and affect pilots of high-altitude aircraft such as the U–2 and SR–71. Sorties could be rerouted or rescheduled to avoid exposing the crews to excessive radiation. NASA used these data in rescheduling Space Shuttle flights.

Navigation Satellites-The Navstar Global Positioning System

In the early 1970s, the Air Force acquired responsibility for an additional space mission—navigation satellites. The first navigation satellite system, the Navy's Transit program, began launches in 1960 and the system became operational in 1964. Transit used precise radio signals from satellites in known orbits, timed with on board atomic clocks, to provide position data to Navy ships and submarines. If the speed of the ship was known exactly, its location could be determined with an accuracy of 200 feet in latitude and longitude. (An error of 1/2 knot reduced this to 600

A GPS satellite undergoing checkout. The GPS system was critical to the success of Coalition forces in the 1991 Gulf War. Although built as a military system, its civilian uses are now widespread.

feet.) Although Transit was widely used by the Navy and commercial shipping, it could not be used by an aircraft moving at high speed, nor could it give altitude measurements. An improvement in accuracy to tens of feet, rather than hundreds, would also greatly increase the system's usefulness. In 1964, both the Air Force and Navy began studies of improved navigation satellites.

The Air Force concept, called 621B, envisioned 20 satellites in geosynchronous orbits. The system could provide world-wide, three-dimensional (latitude, longitude, and altitude) coverage. The Navy system, called Timation (short for Time Navigation), used 21 to 27 satellites in eight-hour orbits. Like 621B, the Navy satellites also would carry atomic clocks to provide the time signals. Two Timation satellites were launched in 1967 and 1969, while tests of 621B were conducted at the White Sands Proving Grounds, New Mexico, in 1971 and 1972.

In 1973 Defense Department officials evaluated these two programs and combined them. Future U.S. military navigation satellites would use the 621B's signal structure and format, but be placed in slightly higher orbits like those planned for Timation, with a 12-hour rather than an eight-hour period. The Defense Department identified the Air Force as the lead service to oversee development and operation, and named the program the Navstar Global Positioning System (GPS). North American Rockwell was selected as prime contractor to build the instrumented GPS satellites,

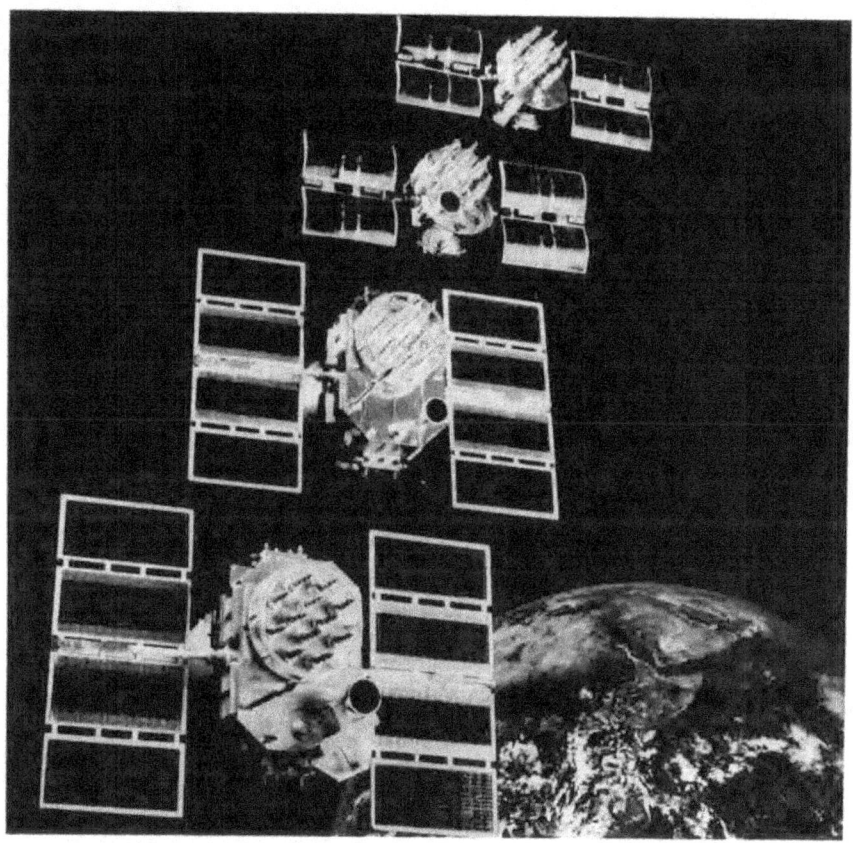

The design evolution of the Global Positioning System satellites from the
early test satellites (top) to the most recent GPS version (bottom).

which would be launched atop Atlas E/F boosters. GPS represented a
national system intended to serve the military needs of the Air Force,
Navy, and the Army. As events transpired, it also would bring about a revolution in commercial transportation management, surveying, and air-sea
rescue.

The original GPS plan called for a network of 21 satellites (18 primary and three spares), later expanded to 24 satellites (21 primary and
three spares). The GPS satellites were placed into 10,900-nautical-mile
circular orbits, inclined 55 degrees to the equator. The full complement of
satellites, located in six orbital planes, insured that four satellites were visible to a terrestrial receiver at all times. Each GPS satellite carried four
atomic clocks and a low-power transmitter which broadcast coded time
signals and status messages. By integrating the signals from four satellites, a GPS receiver could determine its location, altitude, speed, and time
with extreme precision. Nonetheless, to prevent the enemy for using GPS,
it was designed to provide two levels of accuracy in position. In military

use, it transmitted an encrypted signal, the P(Y) code, allowing an accuracy of 52 feet. The key for the P(Y) code was changed every day. In civilian use, commercial receivers worked from an unencryped signal that provided an accuracy in position of 247 feet. If any receiver remained stationary in the same spot for several hours, however, with repeated integrations much higher accuracies became possible; for example, in the 1990s GPS transformed the business of surveying properties and of monitoring the movement of earth along geologic faults.

The Air Force launched the first three GPS satellites in February, May and October 1978. This allowed tests using a satellite/ground transmitter mix located at the Yuma Proving Ground in Arizona. With the launch of the fourth GPS in December 1978, full-scale testing began. With its 12-hour orbits, this constellation of GPS satellites were in sight of the Yuma Proving Ground for about two hours per day. Tests using aircraft, helicopters, trucks, jeeps, and even 25-pound backpack receivers were conducted. Subsequent tests with ships at sea were also successfully made. The Defense Department approved full-scale development of the GPS system in June 1979.

GPS made possible a tremendous advance in field operations for all three military services. For example, since the early 18th century soldiers depended on an available maps and compasses for land navigation. These often proved nearly useless at night, in a jungle, or in a trackless desert. More important, directing artillery fire or calling in close air support required higher accuracies than a compass, map, and wristwatch could provide. In the Yuma GPS tests, Army forces attained accuracies better than 30 feet in all three dimensions, and 0.1 feet per second in speed were achieved in any weather, night or day. The Air Force launched the fifth GPS satellite in February 1980. But, in all, only 11 Block I GPS research and development satellites were launched between 1978 and 1985. The later Block I GPS satellites also carried Vela nuclear detection instruments, taking over some of that mission from other military satellites. Launches of the operational Block II and IIA satellites began in 1989.

Although originally intended to become fully operational by the late 1980s, the military services delayed funding this space system, favoring instead other terrestrial commitments. During the Gulf War in late 1990 and early 1991, only fifteen GPS satellites were operating on orbit. Worse for the military services, they had neglected to purchase military receivers for most of their aircraft and surface vehicles. Orders were hurriedly rushed for commercial ones, and the GPS encryption had to be turned off to permit their use. Nevertheless, the immense advantage that GPS navigation data provided properly equipped Coalition surface and air forces did convince most, if not all, American military leaders to purchase and install military receivers on their fighting vehicles. In the war's aftermath, the Air Force also moved quickly to launch the remaining satellites needed for the GPS system to reach its full complement of 24. The Defense

Department declared the Navstar GPS satellite system fully operational on July 17, 1995.

Anti-Satellite Systems-SAINT, 437, and the F–15 ASAT

Part of the public's emotional response to the launches of the early Sputniks in 1957-1958 reflected fears that the Soviets would launch nuclear weapons into Earth orbit that could be called down on American heads at a moment's notice. President Eisenhower's scientists sought to explain away these public concerns as best they could. Compared with nuclear-tipped ICBMs that could traverse the globe in 30 minutes to strike a target, orbiting nuclear weapons in space not only would cost enormous sums more, they said, but such weapons made no technical or military sense. First, an orbiting weapon required elaborate spacecraft systems, such as retro-rockets to deorbit it, others to guide it, and still others to arm it. Second, all of these integrated systems would have to perform reliably while on orbit for many months if not years, or the bomb became useless. (Spacecraft at that time hardly performed reliably for more than a few orbits, as the MIDAS experience made clear.) Third, if used in retaliation, such weapons could not be delivered at a moments notice, but would have to wait at least an orbit or two until the Earth turned beneath it and the intended target hove into view. Finally, and perhaps most tellingly, if such a weapon were used for a first strike and a partial malfunction occurred as the nuclear bomb moved along its orbit, it might just as easily fall on Buenos Aires as on Washington D.C., or, worse yet, on Moscow. Consequently, the Eisenhower Administration disdained serious military efforts to counter such weapons.

In August 1959, the Air Force Ballistic Missile Division issued a preliminary development plan for a Satellite Interceptor (SAINT). In mid-June 1960, Herbert York, Director of Defense Research and Engineering, ordered the SAINT program restricted to development of subsystems and forbade flight testing. In a review of the revised SAINT plan in July 1960, Air Force Undersecretary Joseph Charyk, involved at that time in evaluating the best way to organize the Air Force reconnaissance satellite effort, ordered any reference to a satellite "kill" capability removed from the SAINT program. He restricted SAINT to rendezvous and inspection only, which brought with it a new name, Satellite Inspector. The Eisenhower Administration's opposition to a weapons-equipped satellite interceptor reflected concerns about jeopardizing the principle of "Freedom of Space"; officials sought to ensure and preserve the right of unobstructed passage in space for reconnaissance satellites. Space-based military weapon systems, on the other hand, might be judged legitimate targets for destruction—an altogether different precedent that U.S. leaders were unwilling to encourage.

The public statements of Soviet officials in the early 1960s occasionally implied that orbital nuclear weapons might be militarily useful. At

Johnston Island was the site for the Program 437 satellite interceptor between 1964 and 1975. Program 437 used a Thor booster and a nuclear warhead.

least they exploited the prospect of them through other space activities in an effort to intimidate the West. When John F. Kennedy succeeded Dwight D. Eisenhower as President of the United States in 1961, the issue of nuclear weapons in orbit at first seemed open to question in the new administration. In June 1962, Harold Brown, who replaced Herbert York as Director of Defense Research and Engineering, rejected an expansion of the SAINT program, saying that "this program should proceed only at an orderly pace on a strictly R&D basis." At the same time, Secretary of Defense Robert McNamara called on the Air Force to "get on with the SAINT program." If not intended to counter orbiting nuclear weapons, the Soviets already had made threats against U.S. reconnaissance satellites. There was a need, McNamara said, for the U.S. to be able to warn its protagonists: "If you shoot down one of ours we will shoot down one of yours." But if one had to deal with a military space threat, another answer seemed to lie in Earth-based anti-satellite weapons.

A few weeks before, in May 1962, McNamara had approved testing of the Army's Nike Zeus Anti-Ballistic Missiles (ABMs) as anti-satellite interceptors. Unarmed tests were conducted from White Sands Missile Range in New Mexico, then from Kwajalein Atoll in the Western Pacific. The Nike Zeus satellite interceptor (called Program 505), however, was limited to a maximum altitude of about 200 miles. On September 12, Air Force leaders submitted to Air Force Secretary Eugene Zuckert a preliminary plan that used Thor IRBMs as satellite interceptors. This larger rocket provided a much greater interception capability than did the Nike Zeus. The nuclear-tipped Thor missiles would be based on Johnston Island in the Pacific and use facilities built for the 1962 high-altitude Fishbowl nuclear tests. Zuckert was critical of employing a nuclear warhead, but ordered a more complete plan be submitted to him as soon as possible.

The Cuban Missile Crisis of October 1962 moved U.S. anti-satellite programs off "top dead center." Previously, CIA analysts and White

House officials had judged Soviet deployment of nuclear-armed missiles in Cuba as "irrational." Yet, Khrushchev had done so and brought the world to the brink of nuclear war. Might the Soviets also be so irrational as to launch orbital nuclear weapons? A State Department contingency plan prepared in May 1963 observed:

> It is possible that the Soviets may at any time conclude that the politico-military and psychological gains from such a feat would justify its undertaking . . . . A thermonuclear "sword of Damocles" would seem to hang over everyone's head in a way which, logic and military technology aside, ICBMs do not.

The plan concluded:

> In anticipation of the contingency of a Soviet weapon in space and recognizing that it may be necessary to undertake physical countermeasures, we should develop as rapidly as possible anti-satellite capabilities.

By February 1963, the Thor interceptor (now named Program 437) was judged the best way to gain greater altitude capability over that offered by Program 505. Zuckert told the Chief of Staff "that development of an operational capability to negate satellites has top priority among defense programs." Program 505 would go into operation as an interim experimental system. (Owing to its limited altitude capability, it would be closed down in 1966.) Work also began on Program 437, with the aim of having the Thor interceptors on 24-hour alert.

Although this action contrasted sharply with the cautious approach accorded SAINT (which Secretary McNamara had cancelled in December 1962), doubts remained about anti-satellite programs in general. In late 1963, Kennedy Administration officials met to review the technical feasibility and political sensitivity of Program 437. In attendance: Robert McNamara, Harold Brown, Undersecretary of State U. Alexis Johnson, Director of the U.S. Information Agency Edward R. Murrow, and Col. Harry E. Evans, Chief of the Research and Development Division of the Joint Chiefs of Staff. In the official Program 437 history, Evans recalled:

> Most of the civilian leadership of both the State and Defense Departments were very nervous about even having a program of research and development for something like 437, let alone the prospect of having such a system operationally ready and manned by "blue suiters." Certainly the aspect of detonating a nuclear weapon in space was politically unattractive to them.

As the discussion continued, most seemed to be against what was viewed as a political liability. Up to this point, Murrow had been quietly smoking a cigarette. Now he interrupted the discussion with a brief comment:

> If the Soviets place a bomb in orbit and threaten us and if this administration has refused to develop a capability to destroy it in orbit, you will see the first impeachment proceeding of an American President since Andrew Johnson.

Evans recalled that about two minutes of total silence followed Murrow's remark. Finally, McNamara said testily, "Well, it doesn't cost much, and the JCS want it, so let's approve 437."

Program 437 featured a novel operational profile. The Thor could intercept a satellite when it passed over Johnston Island at altitudes as high as 700 nautical miles, and within a cross range of 1,500 nautical miles. Because of timing requirements, the missile had a launch window of plus or minus *one second*. Two Thors would be counted down simultaneously—one the primary interceptor, the other serving as back-up. The Thor would lift off and follow a ballistic trajectory to the intercept point. A radio signal would then arm and detonate a MK 49 nuclear warhead. With a yield of one megaton, it had a kill radius of five miles.

Because Program 437 used existing equipment and enjoyed a DX priority, development went smoothly. The first test launch took place on the night of February 14, 1964. The simulated warhead passed within the kill radius of the target satellite, a Transit 2A rocket body, and the interception was judged a success. A second test launch on March 1 went less smoothly. The primary missile was "scrubbed" because of instrument problems. The countdown continued with the back-up missile, and it was successfully launched. Although not perfect, the launch had shown Program 437 could cope with problems during the countdown and still make a successful interception.

The first two Program 437 launches completed nearly all the Phase I test objectives. The Air Force decided to bring the system to full operational (Phase II) status. The third launch was conducted successfully on April 21, 1964. The fourth test launch was cancelled, then rescheduled for mid-May as an all-Air Force operation (called a Combat Training Launch or CTL). This first Air Force CTL on May 28, 1964 proved less fortunate. The Thor lifted off successfully but a booster guidance failure caused it to miss the intercept point. Nevertheless, the next day, on May 29, 1964, Program 437 was judged to have achieved Initial Operating Capability with a single Thor. On June 10, the system was declared fully operational when a second Thor was placed on alert. President Lyndon Johnson made public the existence of the Nike Zeus and Thor anti-satellite weapons on September 20, 1964, during a campaign speech.

Although Program 437 might be operational, events were already limiting its usefulness. The original plan called for Combat Crews A, B, and C to each make one Combat Test Launch per year. In December 1963, the Defense Department notified the Air Defense Command that the number

62

of Thors allocated to the program was being cut from sixteen to eight. As four were needed to keep Program 437 operational (two on alert at Johnston Island and two in war reserve at Vandenberg AFB), this left only four Thors for CTLs through FY 1967. The first CTL was conducted on November 16, 1964, and the second on April 5, 1965. Both were successful. But it would be nearly two years before another one took place. In the meantime, Program 437 began to shift away from being an operational anti-satellite system, its resources diverted to support an experimental satellite inspector.

Early in 1964 work began on adapting Program 437 to undertake the inspection role originally planned for SAINT. Initially called Program 437X, it was re-named Program 437AP (Alternate Payload). The nuclear warhead was replaced with a camera payload. The inspection package could photograph a satellite at altitudes between 100 and about 400 nautical miles, and out to a maximum slant range from Johnston Island of 800 nautical miles. Once the photo run was completed, the exposed film would be taken up on cassettes within a recovery capsule, which would be ejected as the payload re-entered the atmosphere. The capsule would then deploy a parachute and be retrieved in mid-air by a modified C–130 transport.

The first Program 437AP test launch was made on December 7, 1965, with an Agena satellite body serving as the target. The lift-off and photo pass was normal, but during reentry a short circuit prevented capsule separation, and it was destroyed. The two subsequent Program 437AP test launches, on January 18 and March 12, 1966, were successful. Col. Merle M. Zeine, the program director, recommended canceling the planned fourth test launch and using the payload for an operational mission. Both Air Force Systems Command and the Air Staff approved. The NORAD commander requested from the Defense Department permission to conduct an operational mission on April 6 to photograph a Soviet satellite. By this time, however, political sensitivity over even inspecting a Soviet satellite and the chance of an international incident, caused permission to be denied. (The Program 437 and 437AP test launches, as well as the CTLs, had all been directed against inactive U.S. satellites.)

Interest in using Program 437AP launches against Soviet satellites remained widespread among operational commanders in the field and those in the intelligence community. In late April 1966, Air Defense Command and Air Force Systems Command tentatively agreed to make ten Program 437AP launches between 1967 and 1969 in a program called "Stone Marten." Harold Brown, now Secretary of the Air Force, wrote in a memo to McNamara that the intelligence community had requested an enhanced capability to gather data on Soviet satellites, an area where current capability was considered poor, and forecast "the distinct possibility of a future requirement for its use."

While future Stone Marten launches were debated, a NASA satellite served as the target for the first operational Program 437AP launch. On

April 8, 1966, the Orbiting Astronomical Observatory-I was launched into a nearly circular 500-nautical mile orbit. Two days later, OAO-I suffered a power system failure which crippled it, and NASA officials requested that the fourth Program 437AP launch be used to photograph the satellite. On April 17, the Air Staff agreed. The flight would test the altitude and range limits of Program 437AP. The launch from Johnston Island was made on July 2, but a short circuit caused it to diverge from the planned flight path. The camera could not acquire OAO-I, and it photographed empty space.

The failure of the OAO-I mission ended Program 437AP. The United States Intelligence Board considered further launches from Johnston Island to be inadvisable. The site had been publicly identified as an anti-satellite base, and the Soviets were sure to track American rockets and discover if one of their satellites had been photographed. The Intelligence Board suggested that a base in the continental United States be used, but the cost estimates for such an effort made it prohibitively expensive. On November 30, 1966, Air Force Headquarters formally cancelled Program 437AP.

Program 437 now returned to full alert status. The first CTL in nearly two years was conducted on March 31, 1967, as part of Air Defense Command exercise which brought Program 437 into simulated wartime operations for the first time. A simulated warhead passed within two nautical miles of its target, a piece of space debris. (The shortage of missiles which had limited CTLs was eased when an additional nine Thors were allotted by Defense Department to Program 437 CTLs through 1971.) Another successful CTL followed on May 14, 1968. The simulated warhead's miss distance was one-and-one-quarter nautical miles, against an Agena satellite. The CTL was conducted as part of a Joint Chiefs of Staff and National Military Command Center exercise, which gave the training launch an air of realism. The second CTL of 1968, conducted on November 20, was also successful.

Despite Edward R. Murrow's admonition and still without any Soviet weapons to attack in orbit, in 1969, Program 437 began to be dismantled. Moreover, one year earlier the Soviet Union and the United States had signed the UN-sponsored Outer Space Treaty, which prohibited stationing "weapons of mass destruction" in space. That particular threat appeared to have all but disappeared. The number of personnel assigned to the project was reduced, which required that nuclear warheads be removed from their Thor missiles and placed in secure storage elsewhere on the island. The need to remount the warheads increased the reaction time from five to eleven hours. Before year's end, the Defense Department announced phaseout of the complete system by the end of FY 1973.

A final Program 437 CTL was conducted on March 27, 1970, and a simulated warhead passed only one nautical mile from the target satellite. Two months later, on May 4, 1970, Deputy Secretary of Defense David Packard directed the Air Force to accelerate phase down of Program 437 to standby status by the end of the fiscal year. There was, he said, little likelihood that the system ever would be needed. With its 24-hour alert status cancelled, the program's missiles and warheads were removed from Johnston Island and the launch and ground facilities shut down. Now it would require 30 days to return Program 437 to operation. If a final fillip were needed, Hurricane Celeste delivered it on August 19, 1972. High winds and tides struck the island and damaged the program computers and other facilities. Local efforts at repair failed, and Program 437 was removed from service on December 8, 1972. Not until March 20, 1973 was the damage repaired and the program returned to stand-by 30-day alert status. Although the practical ability of Program 437 to destroy a Soviet orbital weapon was now minimal, it still represented the only U.S. anti-satellite system. For this reason, it was retained temporarily. The Defense Department finally terminated Program 437 on April 1, 1975. The order, issued on an inauspicious date for program participants, brought to a close the first operational Air Force anti-satellite program.

Although the Program 437 satellite interceptor program had been reduced to a "stand-by status" in 1970, Defense Department and Air Force interest in anti-satellite systems continued. The Soviets, meanwhile, had developed and tested an anti-satellite (ASAT) interceptor, similar in some respects to the SAINT concept, which relied on a high-explosive warhead to destroy a target satellite. Without a counter, Air Force officials argued, the Soviets now could destroy U.S. satellites without any risk of retaliation in kind. By the early 1970s, the Air Force had begun to reconsider an air-launched ASAT. This concept, which dated back to the late 1950s, drew on attempts to intercept a satellite with an air-launched rocket. During September 1959, a missile was launched from a B–58 in an attempt to intercept the Discoverer 5 satellite. The launch ended in failure. On October 13, 1959, a Bold Orion missile was launched from a B–47 and passed within four miles of Explorer 6. This represented the first interception of a satellite.

In 1971, the air-launched ASAT concept re-emerged with a proposal for Project SPIKE. The concept featured an F–106 interceptor carrying a Standard Anti-radar Homing Missile with a small second stage and a terminal homing vehicle. The ensemble incorporated a target seeker, horizon sensors and an on-board computer. The payload could be either a small nuclear or high-explosive warhead, or a photographic package. Although SPIKE was not developed, it established the basic design features that would be used.

In 1975, President Gerald Ford approved the beginning efforts on an Earth-based F–15-launched ASAT. The anti-satellite rocket in this instance was about 18 feet long and 20 inches in diameter. Five control fins provided guidance and stability during launch within the atmosphere. The solid-propellant rocket's first stage consisted of a modified SRAM missile, while the second stage was an Altair III rocket from the Scout

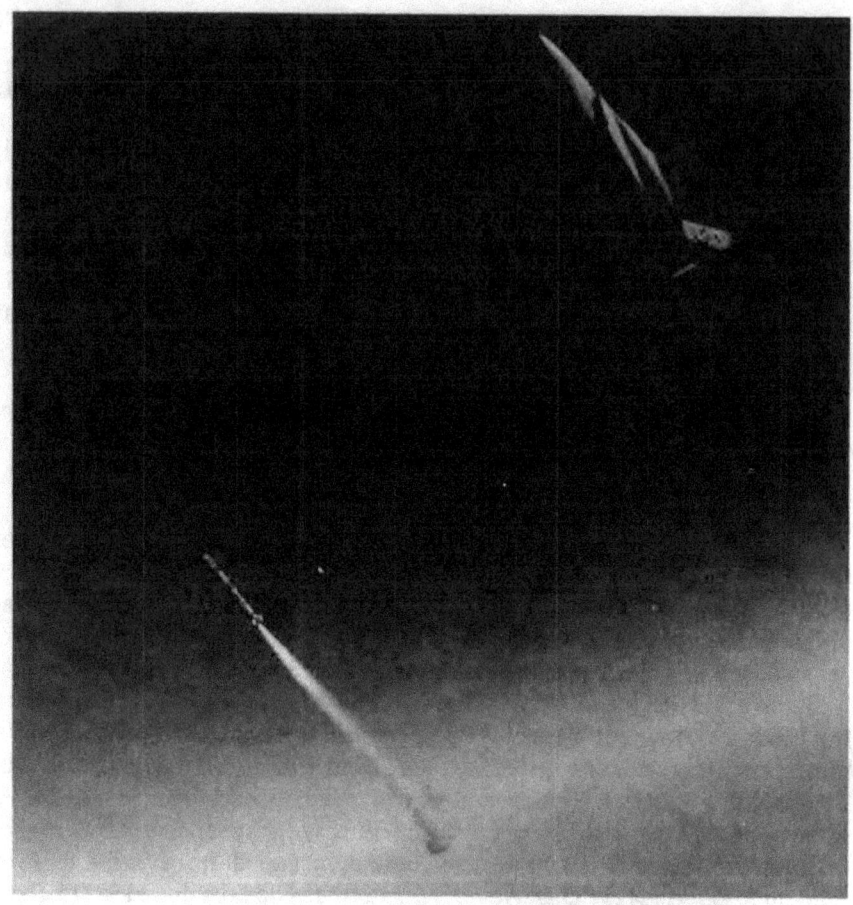

Major Doug Pearson launches the ASAT on its way to intercept and
destroy the Solwind satellite.

booster. The missile carried a "miniature vehicle" which would make the
interception. This ASAT missile would be released from the F–15 while
in a steep climb. After the first stage burned out, it would separate and the
second stage would ignite. The guidance system would direct the second
stage to the intercept point. After second stage burnout, the miniature
vehicle would be released and steered to collide with the target satellite by
64 small solid-fuel thrusters. The high-speed impact would blast the target satellite into a cloud of fragments.

On May 26, 1983, the first of 13 all-up captive flights tested the ASAT
missile's ability to navigate accurately, the F–15's ability to fly the launch
profile, and the ability of the range instrumentation to monitor the launch
program. The first launch of the ASAT missile on January 21, 1984, tested
the missile's ability to fly to a point in space and deploy its miniature vehicle, although a miniature vehicle was not carried in this test. On

November 13, 1984, the first test of the miniature vehicle, using a star as the target, proved unsuccessful. The third launch represented the first (and only) actual interception. A fully-developed ASAT missile was launched on September 13, 1985, and the miniature vehicle struck and destroyed the P78-1 Solwind satellite as it orbited the Earth at an altitude of 320 miles. Loss of the satellite's telemetry signals marked the moment of impact. The final two tests, on August 22 and September 29, 1986, were directed toward stars.

Originally, the Air Force sought to equip two F–15 squadrons with ASAT missiles in 1988 stationed at Langley AFB in Virginia, and McCord AFB in Washington. But the program stirred controversy in Congress, where it was viewed by many as the start of a space arms race. Congress imposed restrictions on space testing (the one interception was conducted between the lapse of one Congressional testing ban and the passage of another). Test restrictions and budgetary limitations caused Air Force leaders to cancel the program in March 1988. The Congressional debate over the Earth-based F–15 ASAT, however, represented only a small part of a still-larger debate over space-based weapons.

The Strategic Defense Initiative

In the mid-1970s, about the time the F–15-launched ASAT program started, political and military interest increased in Anti Ballistic Missile (ABM) systems that might protect the United States from a missile attack in the event of war. Soviet ICBM forces had grown from a handful of missiles in the early 1960s to 1,500 by the early 1970s, when the SALT I Treaty limited the number of U.S. and Soviet missiles that could be deployed. By the end of the 1970s when the Soviets invaded Afghanistan, many Americans believed the U.S. was in a military decline, while the Soviets were aggressively expanding their influence. In 1980 Ronald Reagan was elected President, pledging to increase U.S. military strength. In 1983 he directed the start of the Strategic Defense Initiative (SDI), a wide-ranging research and development program that could provide Earth- and space-based ABM systems, and, it was hoped, shield the United States from attack by ballistic missiles. The effort was managed by the Strategic Defense Initiative Organization (SDIO) attached to the Office of the Secretary of Defense. The Air Force had a major role in SDI's study and test programs.

SDI included improved early warning systems, traditional Earth-based ABM missiles without nuclear warheads (with destruction of a target secured through direct impact), and constellations of Earth-orbiting battle stations. The latter would carry missiles on board designed to home on and destroy Soviet ICBMs during their boost and mid-course phases. In the longer term, SDI looked at directed energy weapons which included Earth-based, airborne, and orbital lasers, as well as particle beam weapons such as the X-ray laser.

In August 1987, the Defense Acquisition Board selected three Air Force SDI projects for further development. Two of these were surveillance systems that involved detection of missile launches and mid-course correction phases. They were called the Boost Surveillance and Tracking System (BSTS) and the Space Surveillance and Tracking System (SSTS). In August-September 1990, BSTS was transferred from SDIO to Air Force control, and was re-oriented towards a more narrow mission of tactical warning and attack assessment. Plans called for it to replace the existing DSP early warning satellites. The program was subsequently renamed the Advanced Warning System (AWS) and later the Follow-on Early Warning System (FEWS). The SSTS program remained under SDIO control, but was subject to several changes in structure and concept. Flight experiments were cancelled, the planned number of satellites scaled back, and the program's cost reduced. In July 1990, this program was renamed "Brilliant Eyes."

The third program, called the Space-Based Interceptor (SBI), was designed to consist of hundreds of Earth-orbiting satellite battle stations, each one equipped with small rockets to destroy Soviet ICBMs. By 1990, SBI had evolved into the "Brilliant Pebbles" concept. This employed a similarly large number of highly autonomous satellites, but with their own targeting systems, reducing the need for separate tracking systems like BSTS. Hardware tests of the Brilliant Pebbles interceptor were conducted in a hangar at Edwards AFB in California. During these ground tests, the satellite supported itself in midair with a rocket engine while other rocket engines pointed it towards a rocket exhaust plume some distance away. These tests demonstrated the ability of Brilliant Pebbles' sensors and computer systems to seek out and lock on a target.

These three systems formed the basis for the first phase of SDI deployment. Looking to the future, ground and airborne tests of high-power lasers also were conducted, and laboratory tests of particle beam weapons were made. Over time, these SDI directed energy weapons programs were reduced in scope and their emphasis shifted to technology development. Needless to say, because of the enormous costs projected to develop and deploy these weapon systems and their ramifications for altering long-standing defense and space policies going back to the 1950s and 1960s, the SDI program proved highly controversial. For example, SDI challenged vital elements of the strategic deterrence policy known as Mutually Assured Destruction. Part of it, the 1972 ABM Treaty between the United States and the Soviet Union, denied to each power more than a limited number of ABM missiles; thus, each side knew that a nuclear first strike would entail certain retaliation in kind and an end to both countries. Even testing, much less deployment, of new ABM systems required changes in the 1972 ABM Treaty. Despite protestations to the contrary from representatives of the Reagan Administration, SDI appeared to many as a nuclear first strike initiative: Once America's orbital battle stations were deployed, this country could attack the Soviet Union in the certain knowledge that a Soviet counterstrike against the United States would fail. Some characterized SDI as "destabilizing," militarily useless, and technically and economically unfeasible. The debate often proved harsh and personal.

Before these issues could be decided, however, in December 1991 the Soviet Union ceased to exist—replaced by a much weaker Commonwealth of Independent States. But other events underscored the political and military significance of ballistic missile defense just a few months before the Soviet Union disappeared into the dust bin of history.

IV

Launch Operations, Ground Control, Organization and Management

If automated defense application satellites are the most visible element of the military space program, many people and numerous organizations support and operate them. All American space launches are conducted from Cape Canaveral, Florida, and Vandenberg AFB, California. The Long Range Proving Ground AFB (later Patrick AFB) at Cape Canaveral was formally established on June 10, 1949 as a guided missile test facility. Over the years that followed, it saw the first launches of the Redstone, Atlas, Titan, Thor, and Jupiter missiles. When the U.S. entered the Space Age, it was the launch site for the early Vanguard and Explorer IGY satellites, Moon and planetary probes, and all U.S. manned space flights. Cape Canaveral, in fact, hosted the launch of all American civil and military spacecraft eastward into low inclination Earth orbits.

Cape Canaveral, however, could not be used to launch satellites into high-inclination orbits, also called "polar orbits." Such orbits would require the rocket booster to overfly populated areas in Florida, Cuba, and South America. Should a failure occur, falling debris might cause injuries or deaths. Clearly, a second U.S. launch site was needed. The site selected on September 1, 1956, was a former Army base called Camp Cooke. On October 4, 1958, Cooke AFB was formally renamed Vandenberg AFB, in memory of Air Force Chief of Staff, General Hoyt S. Vandenberg. The first launch from Vandenberg AFB, of a Thor IRBM, was made on December 16, 1958. Located on the California coast, a booster could be launched to the south and west without crossing land. Any Air Force satellites that required polar orbits, such as SAMOS, the early MIDAS, or DMSPs were launched from Vandenberg AFB.

Air Force crews performed the checkout and launch of all Thor, Atlas, and Titan military boosters. The 6595th Aerospace Test Wing (ATW) conducted all military launches from Vandenberg AFB, while the 6555th ATW had that responsibility at Cape Canaveral. Thus, the 6555th ATW

Onizuka AFB in Sunnyvale, California, served as the hub of the world-wide Air Force spacecraft tracking and control network before 1993. This Air Force network went into operation in 1960 and was transferred to Space Command in October 1987.

conducted the launches of NASA's Mercury and Gemini manned spacecraft that used military boosters. Once a military spacecraft reached orbit, Air Force controllers on the ground had to monitor its operations, send commands to the space vehicle, and receive and distribute data from it. For Department of Defense satellites, the Air Force Satellite Control Network performed this function. An interim satellite control center was originally established in Palo Alto, California, in January 1959. By June 1960, a permanent facility had been built at Sunnyvale, California, near Lockheed's Missile and Space Division, which built the Agena booster-satellite. The Air Force ground control center was originally called the Satellite Test Annex, then, through the 1960s and into the mid-1980s, Sunnyvale Air Force Station (AFS). A less formal name, the "Blue Cube," described the large, multi-story, windowless main building.

Data from the military satellites fed into the Blue Cube via a network of tracking stations. The original nine sites, built between 1959 and 1961, were scattered across the world, from the Seychelles Islands in the Indian Ocean to Thule Air Base in Greenland. Over the years some of the sites were closed, while others were added. (By the early 1980s, a total of seven

The ground tracking station at Vandenberg AFB. A network of such
Air Force ground stations track, command, and control all U.S.
military satellites as they orbit the Earth.

were in operation.) Controlling the military satellites was a continuous
process. Preparations began about 30 minutes before a satellite passed
over a particular tracking station. Once the pass began, the tracking
antenna would follow the satellite and receive its telemetry signals. When
military communications satellites became operational, data from other
defense application satellites could be relayed through communication
satellites to Sunnyvale AFS in near real time. The data was recorded, then
reviewed, to detect any problems. Commands to the satellites would also
be transmitted during a pass. These ranged from routine "housekeeping"
messages to emergency commands to deal with a subsystem failure. It was
much easier to control satellites in high orbits, such as those at geosynchronous altitude, than low-orbit satellites (90 to 450 nautical miles). The
latter required "great care and feeding" owing to the short time a satellite
remained within view of a ground station.

Management of Air Force space activities underwent a series of
changes during the 1950s and 1960s. The first Air Force organization
responsible for space activity was the Western Development Division,
established in July 1954. At first, WDD was charged only with the Atlas
ICBM. Within eighteen months, its responsibility grew to include the
Titan ICBM, the Thor IRBM, and the WS-117L reconnaissance satellite.
On June 1, 1957, WDD was renamed the Air Force Ballistic Missile
Division (AFBMD) which reflected its expanded role.

As space activities became a larger part of AFBMD's activities, in
April 1961 the organization was replaced by the Ballistic Systems
Division (BSD) and the Space Systems Division (SSD). This was part of
a wide-ranging change in the Air Force's research and development orga-

nization. The former Air Research and Development Command became the Air Force Systems Command, while the Air Materiel Command became the Air Force Logistics Command. This first division of missile and space activities proved short-lived, and in July 1967, BSD and SSD were reconsolidated to form the Space and Missile Systems Organization (SAMSO), which continued to operate until the late 1970s.

In 1960 the Air Force established the Aerospace Corporation as a non-profit institution (like RAND) to perform systems engineering for and technical direction of Air Force missile and space contractors. Aerospace Corporation technical personnel were involved with nearly every military satellite and missile program. For example, the early MOL experiment proposals were prepared by the Aerospace Corporation, as were evaluations of the MOL contractor proposals and cost estimates.

The Founding of Space Command

Air Force space activities grew tremendously in scope and importance between 1958 and 1980. But because no separate operating command for space projects existed, they remained under the control of the service's research and development arm, Air Force Systems Command. After a period of study, on September 1, 1982, Headquarters USAF established Air Force Space Command at Peterson AFB in Colorado Springs, Colorado, to exercise operational control of space systems. The first two satellite systems assigned to Space Command were the Defense Support Program missile early warning network and the Defense Meteorological Satellite Program. In May 1983, Space Command assumed control of all ground-based early warning radars, and, in January 1984, of the Navstar GPS as well. (The fourth defense support space mission, communication satellites, remained the responsibility of the Defense Information Systems Agency.) A few years later, in October 1987, Space Command acquired responsibility the Sunnyvale, California, satellite control center, renamed Onizuka AFB in honor of Lt. Col. Ellison Onizuka, an STS-51C crew member killed in the *Challenger* explosion. Beginning in 1989, the Consolidated Space Operations Center (CSOC) located at Falcon AFB near Colorado Springs, gradually assumed command and control of military satellites in orbit. It became fully operational and was turned over to Space Command in September 1993. At that time, Onizuka AFB became a backup control center. Finally, Headquarters USAF transferred all satellite launch operations from Air Force Systems Command to Air Force Space Command in October 1990.

While Air Force Space Command brought together operational functions, a parallel reorganization at Air Force System Command's Space Division combined the research and development activities. In October 1982, the Air Force Space Technology Center (AFSTC, later redesignated the Phillips Laboratory) was established at Kirtland AFB under the control of Space Division. It was composed of three existing units—the Air

Force Weapons Laboratory, the Air Force Geophysics Laboratory, and the Air Force Rocket Propulsion Laboratory. AFSTC combined Air Force space technology efforts. A further re-organization came in March 1989, when Space Division was renamed the Space Systems Division (SSD) and the Ballistic Missile Office became the Ballistic Missile Division (BMD). This lasted only until May 1990, when BMD was assigned to SSD as an organizational element, the third time the space and missile functions had been combined. In July 1992, SSD itself was re-named the Space and Missile System Center (SMC), reflecting its joint space and missile activities though it remained a component of the Air Force's research and development arm.

The founding of the Air Force Space Command capped changes in the Air Force space mission over three decades. Nevertheless, many military leaders in all of the services still viewed the four primary defense support space missions as something outside the "real world" of Air Force or Navy or Army operations. That attitude changed perceptibly in 1991 when these pre-positioned assets in Earth orbit demonstrated forcefully the central role space support now played in military operations.

V

Desert Storm, the Air Force and the Military Space Program in a Changing World

Desert Storm, the military campaign by U.S. and Coalition forces to liberate Kuwait from Iraqi occupation in January-February 1991, was the first major contest of arms in which space systems were fully integrated at all levels of planning and operations. It has been called "the first space war" because communications, navigation, weather, missile early warning, and reconnaissance satellites proved indispensable to the final success of combat operations. Desert Storm defined how wars would be conducted in the future. The high frontier of space provided U.S. and Coalition forces "information dominance" and, with it, the leverage needed to quickly wage and win a modern war. U.S. and Coalition forces communicated with one another in the field and across the world through military communications satellites. The DSCS satellites provided eighty-four percent of the long-haul strategic communications and much of the shorter-range in-theater tactical communications. Hundreds of satellite communications receivers were employed down to the company level.

In this conflict, the Iraqi Army, equipped with short-range Scud ballistic missiles, could strike at targets in Saudi Arabia and Israel. The Scud's flight time from launch to impact was less than seven minutes—far shorter than that of an ICBM. This made warning time very critical, and the DSP early warning satellites played a vital role. Although Space Command had never attempted warning of tactical ballistic missiles before, during the Desert Shield build-up the Iraqis made three practice Scud launches. These tests allowed the "bugs" to be worked out of the DSP in a quick-response mode. When the Iraqis began Scud launches on the second night of the air war, Space Command was ready.

Data on each Scud launch, detected by the DSP satellites within moments of liftoff, was relayed to NORAD/Space Command at Cheyenne Mountain in Colorado. Within five minutes the command confirmed a launch, predicted the impact point, and flashed word via communication satellites to the Middle East. Space Command personnel then turned to their television sets and watched as the air raid sirens began to wail on the other side of the world. Troops hurriedly put on their chemical protection suits, while civilians took refuge. At the same time, pointed by DSP data, Army Patriot missiles roared aloft to intercept the incoming Scuds. It was all very different forty-six years earlier when London and Antwerp were under attack from German V-2s and the missiles struck without warning. The DSP satellites allowed people to take shelter, while the Patriots provided an active defense. Together, they reduced the number of casualties and damage.

If the DSP and communications satellites were applied successfully in tactical operations, by far the most important automated space system employed in Desert Storm was the Navstar Global Positioning System. Like DSP, it too proved vital to military success. Before the Iraqi invasion of Kuwait, the Army had purchased only about 1,000 military GPS receivers (called Small Lightweight GPS Receivers, SLGRs or "sluggers."). As the Desert Shield deployment continued and the demand for sluggers in the field soared, the GPS Program Office made emergency purchases of some 13,000 civilian GPS receivers for use on military vehicles. Moreover, many soldiers, desperate for the navigational advantage that GPS offered, bought their own from civilian electronics stores or received them as gifts rushed from home. The soldiers were able to have them working within a half hour of opening the box.

Because most of the GPS receivers employed in Desert Storm were civilian models and unable to use encrypted signals, for the greater navigational accuracy the Air Force and Defense Department chose to leave GPS signals unencrypted and risk Iraqi forces also using the same GPS signals. That risk was judged acceptable because Coalition forces were fighting a war of movement in unfamiliar territory, while the Iraqis were tied down in fixed positions, and lacked precision guided weapons that could use the GPS data. But the usefulness of the GPS data also was limited to Coalition forces because the complete satellite network had not yet been established in orbit. GPS provided accurate data when four satellites were in view of a receiver, but in 1991 there were seven periods each day up to 40 minutes in duration when fewer than four were in view. During these GPS "sad times," as they were called, Coalition forces had to revert to LORAN or dead reckoning.

GPS guided Air Force and Navy aircraft equipped with receivers to their targets in the air war. Army special forces helicopters navigated to targets behind Iraqi lines using GPS receivers. And precision artillery fire was directed against the enemy using GPS data. The Navy used GPS data to map mine fields accurately in the Persian Gulf, then navigate through them. But perhaps GPS' greatest achievement came in the "100-hour ground war." The Coalition moved 100,000 troops in poor weather 100 miles west along the Iraqi-Saudi border during the so called "Hail Mary" maneuver. This sidestepped Iraqi forces concentrated in the east near the Persian Gulf, on the Saudi/Kuwait border. At the start of the ground war, four armored divisions with some 2,000 vehicles were lined up along a 75-mile-wide front.

In the border breakthrough of the Iraqi berms on February 26, 1991, two divisions operated side-by-side along a 25-mile front. Using GPS, every unit knew both its location and the planned locations of those on its flanks. The SLGRs, which could also be programmed with waypoints, gave direction and distance to the next waypoint and the vehicle's speed. The GPS' atomic clocks also allowed the time of all the SLGRs to be synchronized. Without GPS, it would not have been possible to conduct the Hail Mary sweep effectively. One Army sergeant gave his SLGR the ultimate compliment—"If it could make coffee, I'd marry it." Captured Iraqi soldiers, who had only compasses and maps, could not believe U.S. forces had found their way through the flat, featureless desert of western Iraq. They were even more astonished when they discovered that GPS receivers were not only in the hands of command personnel, but individual soldiers.

Following the Gulf War, the military services began purchasing additional GPS receivers for use in everything from B–1B bombers to trucks, though hardly as fast as some had hoped. In late 1994, for example, the pilot of an Army helicopter not yet equipped with a GPS receiver strayed across the 38th parallel into North Korea and was shot down. Meanwhile, studies have begun of helmet-mounted GPS antenna and of miniature receivers to be carried by American pilots. If shot down, these receivers will guide rescue helicopters to the aviator.

The DMSP meteorological satellites also played an important role in Desert Shield and Desert Storm. During the build-up of Coalition forces in Desert Shield, a new terminal, the Rapid Deployment Imagery Terminal, was quickly procured. Older terminals were also rushed to the Middle East. These display terminals provided nearly all of the weather data used to support the deployment. Coalition air forces also used the DMSP data to plan initial air strikes. When the lights went out in Baghdad during the first night's air attacks in January 1991, a DMSP satellite first spotted it. This represented the first bomb damage assessment of the air attacks on Iraq's electrical system. The air war itself transpired during the worst recorded weather in fourteen years. Half of all sorties were affected by the weather, resulting in cancellations or diversion to other targets.

DMSP photos helped determine weapons loads and selection of targets. In more than 44,000 combat missions, DMSP weather data was estimated to have saved some $250 million in flights that, without it, would have been recalled.

Air Force Lt. Gen. Donald L. Cromer, Commander of Space Systems Division during Desert Storm, observed the marked change in attitude towards military space systems that occurred during this conflict. Before the Gulf War, he observed, "space people used to be pushed off to the side. We had to fight for everything. We had neither understanding nor strong support for all the things that space could do for the Air Force." Afterward, everything changed. "Operation Desert Shield and Desert Storm will be a watershed for recognizing that space is as much a part of the Air Force and the military infrastructure as airplanes, tanks, and the ships . . . ," Cromer said. "All future wars will be planned and executed with that in mind." Army General Carl Steiner, Commander of the XVIII Airborne Corps, put the change in attitude more succinctly: "Space doesn't just help . . . , I cannot go to war [today] without space systems." Desert Storm proved that pre-positioned military space systems furnished Coalition forces an immense advantage, one that made possible a swift victory with an extraordinarily low number of lives lost in combat.

Six months after the end of Desert Storm, the Union of Soviet Socialist Republics disappeared, and, with it after five decades, the central factor in U.S. military planning. What of the Air Force and the military space program in a post-Cold War future? The Air Force and indeed the entire U.S. military establishment began a major reduction in size. In 1990 the Air Force commanded 200 wings; by 1994 this number had declined to 90. Of thirteen major Air Force commands in 1990, only eight remained in 1994. With the importance of satellites to modern warfare established at the turn of the 21st century, however, defense support space systems fared somewhat better. While overall Defense Department funding declined twenty-two percent between 1992 and 1994, the budget for military spacefaring dipped only eleven percent.

Whether directed to communications, reconnaissance, or early warning of missile attack, the Air Force and its contractors in the 1950s defined virtually all defense support space applications in terms of strategic systems. Desert Storm in 1991 found the military services attempting to apply these strategic space systems as best they could to meet the tactical needs of military commanders in the field. Based on the Gulf War experience, the Defense Department in 1992 established the Combined Imagery Office (CIO) to speed the dissemination of reconnaissance images. Air Force Space Command in 1993 established a Space Warfare Center at Falcon AFB in Colorado to conduct the Air Force Tactical Exploitation of National Capabilities (TENCAP) program. The TENCAP "Talon" programs reflect the future shape of Air Force space activities. These include efforts to demonstrate:

1. **Talon Command** (command and control improvements):

Project Shield—enhanced support from DSP satellites for warning and cuing targets for theater missile defenses.

Project Hook—integrating GPS navigation/position data with survival radios for improved search and rescue operations for downed aircrews.

2. **Talon Ready** (mission planning/rehearsal):

Project Scene—rapid processing of theater and tactical imagery from overhead systems to support target planning for precision-guided weapons.

Project Spectrum—relaying and displaying imagery from civilian Earth resources satellites, such as Landsat and SPOT, as well as national systems, to wing and squadron intelligence and mission planners.

Project Stamp—automated mission planning.

3. **Talon Shooter** (Real-time information to and from the cockpit):

Project Sword—relaying information from space systems directly to tactical aircraft in near real-time, via normal communications channels. In the first demonstration, tactical information was passed to an aircraft making an attack on a simulated surface-to-air missile site. The aircraft then fired HARM (High-speed Anti-Radiation Missile) missiles which destroyed the site, even though the aircraft's own systems had not yet detected it.

Project Lance—placing a mini supercomputer on board an aircraft, which would then receive, process, correlate and display information to the aircrew.

Project Zebra—using highly accurate, imagery-derived coordinates for guided weapons based on GPS Rectified Imagery.

4. **Talon Vision:**

Air Defense System Integration—integrating national- and theater-level intelligence to provide a theater commander with a complete picture of the changing battlefield.

5. **Talon Touch** (Space Communications):

Satellite Launch Dispenser Communications (SLDCOM)—a seamless connection from space systems, through the existing theater battle management communications and intelligence systems, down to individual pilots and soldiers in the field.

Other TENCAP elements include Talon Night, support to special operations forces, and Talon Spear, improved training, testing, and exercise capabilities at Nellis AFB.

In Desert Storm, military leaders also faced the threat of ballistic missiles equipped with nuclear, chemical, and biological weapons. Traditional threats of retaliation proved ineffective in preventing Iraqi Scud attacks on Israel. Saddam Hussein ordered missiles launched against

Israel with the presumed intent of provoking retaliation by a country reported to have nuclear weapons. Such unstable behavior was outside the Cold War experience. The CIA has estimated that by the year 2000, at least six Third World countries will possess ballistic missiles of varying ranges and accuracies.

Post-Desert Storm missile defense planning reflected this threat. In January 1991, President George Bush redirected the Strategic Defense Initiative to a more limited goal of defending U.S. territory, forces, and allies against small-scale ballistic missile attacks. This concept, called Global Protection Against Limited Strikes (GPALS), reduced Brilliant Eyes in size while it continued the Brilliant Pebbles program. On November 10, 1993, the Defense Department cancelled the FEWS early warning satellite and replaced it with a less expensive program of early warning satellites, called the Alert Locate and Report Missiles (ALARM) program. In short order, however, ALARM was replaced in December 1994 by a DPS follow-on called the Space-Based Infrared (SBIR) program.

With the end of the Cold War, to save funds and eliminate duplication, closer ties also have been sought between military and civilian space technology—between the world of classified space technology applications and the world of unclassified civilian space science and applications. For example, the low-altitude military DMSP and its civilian weather satellite counterpart will be merged. The Defense Department also is reexamining the issue of leasing circuits in civilian communications satellites, or incorporate on civilian satellites some military features, such as anti-jamming (much as some aircraft belonging to civilian airlines act as a military reserve airlift force).

Adapting military space technology to civilian and scientific use is not a new story. The GPS satellites have been adopted in the commercial world in ways never dreamed of when the program started in the 1970s. GPS receivers have been installed on board commercial airliners. The same GPS receivers also allow airlines to make landing approaches in virtually any weather conditions. The GPS is also used by ships; the U.S. Coast Guard is "correcting" the encrypted GPS signals for broadcast in some areas. Trucking firms and railroads use the GPS to control and track the movement of vehicles. The ability to measure very slight movement allows GPS to be used to measure the stresses building up in earthquake faults. With the reductions in cost produced by the growth of this industry, GPS receivers are even being offered as an option in new cars. You can now find your way through the Los Angeles Freeway system as easily as the U.S. Army found its way across the Iraqi desert. These commercial applications have become so important that attempts likely will be made to remove this critical space system from military control.

Other automated military satellites have civil scientific and commercial applications not at first foreseen. The Vela nuclear detection satellites were the start of X-ray and Gamma-ray astronomy. This was many years before such specialized satellites as NASA's Gamma Ray Observatory satellite were launched. DSP data has been released detailing asteroid airbursts. These exploded in the upper atmosphere with a yield equal to small nuclear weapons. From SDI came the technology for adaptive optics. Originally designed to allow ground-based lasers to compensate for the effects of atmospheric turbulence, it has since been fitted to astronomical telescope mirrors to improve their resolution. A prime example of this union between military technology and civilian use was the Clementine Moon probe, originally designed to test missile detection and tracking sensors under space conditions. Because NASA had long wanted a lunar probe to make a complete survey of the Moon, as a follow-on to the Apollo missions, the two needs combined in Clementine, which in the mid-1990s was placed into lunar orbit, and its detection equipment produced a high-resolution map of the Moon and its mineral content.

Closer to home, in February 1995, President William Clinton declassified the photographs of the Earth taken by Project CORONA satellites between 1960 and 1972. Originally produced for strategic reconnaissance, they could now be used for Earth resource studies. The CORONA photographs, combined with those of NASA's landsat, provided scientists with a record of images of the Earth spanning 35 years. That record now could be examined for a variety of purposes, such as alterations in the use of land and the growth and movement of populations.

A half century has passed since General Arnold and other astronautical pioneers at RAND and in the U.S. Navy first considered the possibilities of a military space program. In the interval, the Air Force developed space systems that all American military services now employ and upon which they have come to depend. United States defense support space systems have become as central to military operations as air, ground, or sea power. As Desert Storm made plain, they permit these terrestrial forces to be wielded in concert, and they multiply their effects on the battlefield so enormously that an enemy without space assets has little hope of prevailing. If the perceptions of military spacefaring in 1945-1946 became realities in the years that followed, the perceptions of astronautics applied in the TENCAP and Talon programs of today portend even greater possibilities for improved military operations tomorrow.

Acknowledgement

I am indebted to the staffs of the Space and Missile Systems Center and Air Force Space Command History Offices for providing crucial documentation, and especially to R. Cargill Hall of the Air Force History Support Office, for his guidance in organizing and writing this history, and for his editorial counsel on numerous drafts.

Suggested Reading

Austerman, Wayne R., *Program 437: The Air Force's First Antisatellite System* (Peterson AFB, CO: Air Force Space Command, Apr 1991).

Brugioni, Dino A., *Eyeball to Eyeball: The Inside Story of the Cuban Missile Crisis* (New York: Random House, 1991).

Cassutt, Michael, "The Manned Space Flight Engineer Programme," *Spaceflight* (Jan 1989): 26-33.

Davies, Merton E. and William R. Harris, *RAND's Role in the Evolution of Balloon and Satellite Observation Systems and Related U.S. Space Technology* (Santa Monica, CA: RAND R-3692-RC, 1988).

Davies, J.E.D., "The Discoverer Programme," *Spaceflight* (Nov 1969): 405-7.

Greer, Kenneth E., "CORONA," in *Studies in Intelligence*, supplement 17 (Spring 1973): 1-37.

Hall, R. Cargill, "The Eisenhower Administration And The Cold War: Framing American Astronautics To Serve National Security," *Prologue* (Spring 1995): 59-72.

_____, "Origins of U.S. Space Policy: Eisenhower, Open Skies, and Freedom of Space," in John Logsdon, ed., *Exploring the Unknown: Selected Documents in the History of the U.S. Civil Space Program* (Wash D.C.: USGPO, NASA SP-4407, 1995), Vol I: 213-29.

Hallion, Richard P., *The Hypersonic Revolution*, Vol I: *From Max Valier to Project PRIME* (Wright-Patterson AFB, OH: Aeronautical Systems Division, 1987)

Houchin, Roy F., "The Diplomatic Demise of Dyna-Soar: The Impact of International and Domestic Political Affairs on the Dyna-Soar X–20 Project, 1957-1963," *Aerospace Historian*, Dec, 1988.

Klass, Philip J., *Secret Sentries in Space* (New York: Random House, 1971)

Killian, James R. Jr., *Sputnik, Scientists, and Eisenhower: A Memoir of the First Special Assistant to the President for Science and Technology* (Cambridge: MIT Press, 1977)

Kutyna, Gen. Donald J., USAF, "The Military Space Program and Desert Storm," *The Space Times* (Sept-Oct 1991): 3-7.

Mark, Hans, *The Space Station: A Personal Journey* (Durham, NC: Duke University Press, 1987).

Neufeld, Jacob, *The Development of Ballistic Missiles in the United States Air Force, 1945-1960* (Wash D.C.: Office of Air Force History, 1990).

Peebles, Curtis, *Battle For Space* (New York: Beauford Books, 1983).

_____, *Guardians: Strategic Reconnaissance Satellites* (Novato, CA.: Presidio Press, 1987).

Perry, Robert L., *Origins of the USAF Space Program, 1945-1956* (Wash D.C.: Air Force Systems Command, Jun 1962).

Portz, Matthew H., *The Aerospace Corporation, Its Work: 1960-1980* (Los Angeles, CA.: Times Mirror Press, 1980).

Time-Life Books, *Electronic Spies* (Alexandria, VA: Time-Life Books, 1991).

Weaver, Col. Richard L., USAF, "Report Of Air Force Research Regarding The 'Roswell Incident'," (Wash D.C: Office of Air Force History, 1994).

Zaloga, Steven J., *Target America: The Soviet Union and the Strategic Arms Race, 1945-1964* (Novato, CA.: Presidio Press, 1993).

☆ U. S. G. P. O. 2003-496-807/95354